Environmental Risk Analysis

Probability Distribution Calculations

Environmental Risk Analysis

Probability Distribution Calculations

Louis Theodore

CRC Press
Taylor & Francis Group
Boca Raton London New York

CRC Press is an imprint of the
Taylor & Francis Group, an **informa** business

CRC Press
Taylor & Francis Group
6000 Broken Sound Parkway NW, Suite 300
Boca Raton, FL 33487-2742

© 2016 by Taylor & Francis Group, LLC
CRC Press is an imprint of Taylor & Francis Group, an Informa business

No claim to original U.S. Government works

Printed on acid-free paper by CPI Group (UK) Ltd, Croydon
Version Date: 20150721

International Standard Book Number-13: 978-1-4987-1439-6 (Hardback)

Visit the Taylor & Francis Web site at
http://www.taylorandfrancis.com

and the CRC Press Web site at
http://www.crcpress.com

To Rita
my woman Friday

Contents

SECTION II   Discrete Probability Distributions

SECTION III   Continuous Probability Distributions

SECTION IV   Applications

Preface

Environmental risk comes into play in countless real-world industrial applications and poses unique challenges for the environmental community. An integral part of this subject area is the role that probability distributions play in risk analysis. Therefore, the author considered writing a book that highlights pragmatic material with problems (and their solutions) as they relate to this topic. This book will hopefully serve as a training tool for those individuals in academia and industry involved with environmental risk. Although the literature is inundated with texts emphasizing traditional environmental topics, the goal of this book is to present this subject of environmental risk by employing a pragmatic approach.

The book contains four sections. Section I provides the material required for solving many risk analysis problems, particularly as they apply to the environmental engineering and science professions. Sections II and III contain nearly 200 illustrative examples concerned with probability distributions. Section IV provides solved problems involving applications in the following industries: Chemicals and references, energy and power, manufacturing and electronics, pharmaceuticals, military and terrorism, travel/aerospace/travel, and nanotechnology. An abbreviated outline of the topics covered can be found in the Contents.

This book is the result of several years of effort by the author. Part of the original draft was developed during 2007–2014 when the author was preparing material for his books *Environmental Health and Hazard Risk Assessment: Principles and Calculations* (2014) and *Probability and Statistics Calculations in Environmental Science* (2007) (CRC Press). The manuscript has also undergone significant revisions during the past 2 years, some of which were based on the experiences gained from an online course at Montana University and Utah State University, as well as from earlier class testing.

It should be noted that the author cannot claim sole authorship to all the essay materials and problems. This book has evolved from a host of sources, including the two aforementioned books as well as other books published by the author; notes, homework problems, and exam problems prepared by several faculty for undergraduate courses offered at Manhattan College; a host of *Theodore Tutorials* published by Theodore Tutorials of East Williston, New York; and, material developed by R. Dupont and K. Ganesan during the presentation of several courses on this topic. Although a bulk of the material is original or taken from sources that the author has been directly involved with, every effort has been made to acknowledge material drawn from other sources.

In the final analysis, the challenge of what to include and what to omit was particularly difficult. However, every attempt was made to offer most, if not all, environmental risk analysis material to individuals at a level that should enable them to better cope with some of the problems encountered in practice. As such, the book was not written for the student planning to pursue advanced degrees; rather, it was primarily written for those individuals who are currently working as practicing environmental engineers and scientists.

It is hoped that this book, which the author believes covers the principles and applications of probability distributions in environmental risk analysis in a thorough and clear manner, will place a useful resource in the hands of academic, industrial, and government personnel. Upon completion of the book, the reader should have acquired not only a working knowledge of the principles of environmental risk analysis and probability distribution but also experience in their application; the reader should also find himself/herself approaching advanced texts, engineering literature, and industrial applications (even unique ones) with more confidence. The author strongly believes that while understanding the basic concepts is of paramount importance, this knowledge may be rendered virtually useless to a technical individual if he or she cannot apply these concepts to real-world situations.

Last but not least, the author believes that this modest work will help the majority of individuals working and studying in the field of environmental engineering or science to obtain a more complete understanding of the role that probability distributions play in environmental risk analysis. If you have come this far and read through the Preface, you have more than just a passing interest in this subject. It is strongly suggested that you take advantage of the material available in this book; it almost certainly will be a worthwhile experience.

Thanks are due to Kelly Behan for proofing the manuscript pages and special thanks are extended to Rita D'Aquino for her technical and editorial assistance.

Louis Theodore
East Williston, New York

Author

Louis Theodore, PhD, received an MChE and EngScD from New York University, New York, and a BChE from The Cooper Union (New York). Over the past 55 years, Dr. Theodore has been a successful educator at Manhattan College (holding the rank of full professor of chemical engineering), graduate program director (raising extensive financial support from local industries), researcher, professional innovator, and communicator in the engineering field. During this period, he was primarily responsible for his program achieving a No. 2 ranking by *U.S. News & World Report*, and particularly successful in placing students in internships, jobs, and graduate schools. He currently serves as a part-time consultant to the U.S. Environmental Protection Agency (USEPA) and Theodore Tutorials, a company he cofounded that specializes in providing training needs to industry, government, and academia.

Dr. Theodore is an internationally recognized lecturer, who has presented nearly 200 courses to industry, government, and technical associations; has served as an after-dinner or luncheon speaker on numerous occasions; and, has appeared on television as a guest commentator and news spokesperson. He has developed training materials and served as the principal moderator and lecturer for USEPA courses on hazardous waste incineration, pollution prevention, health and hazard risk, and air pollution control equipment. He has also consulted for several industrial companies in the field of environmental management and risk assessment, and has served as a consultant and expert witness for USEPA and the U.S. Department of Justice.

Included in Dr. Theodore's 104 textbooks and reference books are *Pollution Prevention* (CRC Press, 2000), *Engineering and Environmental Ethics* (John Wiley & Sons, 2000), *Air Pollution Control Equipment* (Prentice Hall, 1990), *Introduction to Hazardous Waste Incineration*, Second Edition (Wiley Interscience, 2000), section author and editor for *Perry's Chemical Engineers' Handbook* (McGraw-Hill, 2008), *Nanotechnology: Basic Calculations for Engineers and Scientists* (John Wiley & Sons, 2007), *Chemical Engineering: The Essential Reference* (McGraw-Hill, 2013), and the recently acclaimed *Environmental Health and Hazard Risk Assessment: Principles and Calculations* (CRC Press, 2013). Included in the Theodore Tutorials series of 21 texts are 4 tutorials concerned with the professional engineers (PE) exam. In addition, he was recently involved with the development of a nontechnical environmental calendar that serves as a consumer and youth outreach product.

Dr. Theodore is the recipient of the International Air & Waste Management Association's prestigious Ripperton Award that is "presented to an outstanding educator who through example, dedication, and innovation has so inspired students to achieve excellence in their professional endeavors." He was also a recipient of the American Society of Engineering Education (ASEE) AT&T Foundation Award for "excellence in the instruction of engineering students." In 2008, he was honored at Madison Square Garden by National Pro-Am for his contributions to basketball and the youth of America.

Dr. Theodore is a member of Phi Lambda Upsilon, Sigma Xi, Tau Beta Pi, the American Chemical Society, the American Society of Engineering Education, the Royal Hellenic Society, and is a fellow of the International Air & Waste Management Association (AWMA).

Dr. Theodore is certified to referee scholastic basketball games through his membership in IAABO (International Association of Approved Basketball Officials) and is the author of the 2015 book *Basketball Coaching 101.* He previously served on a Presidential Crime Commission under President Gerald Ford and provided testimony as a representative of the pari-mutuel wagerer (horseplayer). He continues to disturb the peace with his tongue-and-cheek column "AS I SEE IT," which is a monthly feature of several Long Island newspapers, plus his newsletter, www.theodorenewsletter.com, which addresses social, economic, political, technical, and sports issues.

Section I

Introduction to Environmental Risk Analysis

The Prince

There is nothing more difficult to take in hand, more perilous to conduct, or more uncertain in its success, than to take the lead in the introduction of a new order of things.

Nicolo Machiavelli (1469–1527)

Lost Horizon

Perhaps the exhaustion of the passions is the beginning of wisdom, if you care to alter the proverb. That also, my son, is the doctrine of Shangri-La.

James Hilton (1900–1954)

This section contains four chapters.

Chapter 1: Introduction to Environmental Risk
Chapter 2: Environmental Risk Analysis
Chapter 3: Introduction to Probability and Statistics
Chapter 4: Introduction to Probability Distributions

The first two chapters address the general subject of environmental risk analysis. The last two chapters briefly explain the probability fundamentals and principles that are employed in environmental risk analysis. Chapter 3 deals exclusively with probability, while Chapter 4 focuses on probability distributions.

1 Introduction to Environmental Risk

INTRODUCTION

The rapid growth and expansion of the chemical and energy industry has been accompanied by not only a spontaneous rise in chemical emissions to the environment but also human, material, and property losses because of fires, explosions, hazardous and toxic spills, equipment failures, other accidents, and business interruptions. Concern over the potential consequences of these massive emissions and catastrophic accidents, particularly at chemical, petrochemical, and utility plants, has sparked interest at both the industrial and regulatory levels in obtaining a better understanding of the main subject of an earlier book: *Environmental Health and Hazard Risk Assessment: Principles and Calculations.*[1] The writing of that "risk" book was undertaken, in part, as a result of this growing concern.

Risk of all types (health risk, hazard risk, individual risk, societal risk, sports risk, etc.) has surged to the forefront of numerous engineering and science areas of interest. Why? A good question. Some of the more obvious reasons include (not in the order of environmental importance) the following:

1. Increased environmental health and safety legislation
2. Accompanying massive regulations
3. Regulatory fines
4. Liability concerns
5. Environmental activists and their organizations
6. Public concerns
7. Skyrocketing health care costs
8. Skyrocketing workers' compensation costs
9. Codes of ethics

These factors, individually or in toto, have created a need for engineers and scientists to develop a proficiency in environmental risk and environmental risk-related topics. This need, in turn, gave rise to the driving force that led to the writing of this book.

Members of society are confronted with risks on a daily basis. Here is a sampling of some activities for which risk can potentially play a role:

1. Electrocution when turning on the TV
2. Using soap with chemical additives
3. Tripping down the stairs
4. Drinking Starbucks coffee
5. Indoor air pollution

3

6. Driving to work
7. Eating a hot dog for lunch
8. Living downwind from a refinery
9. Being struck by an automobile while returning from lunch

Risks abound. They are all around and society has little to no control over many of them. Perhaps a careful analysis of risks is indeed in order.

Health problems and accidents can also occur in many ways other than from routine, daily, and "normal" activities. As well noted earlier, there may be a chemical spill, a round-the-clock emission from a power plant, an explosion, a terrorist attack or a runaway reaction in a nuclear plant. There are also potential risks and accidents in the transport of people and materials such as trucks overturning, trains derailing, and ships capsizing. There are "acts of God" such as earthquakes, tsunamis, and tropical storms. It is painfully clear that health and hazard problems are a fact of life. The one common thread through all of these situations is that these problems are rarely understood and, unfortunately, they are frequently mismanaged.

The job of the engineer and scientist is to measure or calculate the magnitude of risk and often compare the magnitude of one risk to other risks that are similar in nature. Perhaps more difficult is the task of comparing the risk of one event with risks arising from events of a totally different nature.

Risk Variables

Placing the risk in perspective entails translating myriad technical risk analyses into concepts of risk that both the technical community and the general public can understand. The most effective technique for presenting risks in perspective is to contrast risks to other, similar risks.

There are several variables that affect acceptance of risk. Ten such variables include the following:

1. Voluntary vs. Involuntary
2. Delayed vs. Immediate
3. Natural vs. Man-Made
4. Controllable vs. Uncontrollable
5. Known vs. Unknown
6. Ordinary vs. Catastrophic
7. Chronic vs. Acute
8. Necessary vs. Luxury
9. Occasional vs. Continuous
10. Old vs. New

The public generally accepts voluntarily assumed risk more easily than an involuntarily imposed risk. Similarly, a naturally occurring risk is more easily accepted than a man-made risk. The more similar risks are with regard to these variables, the more meaningful it is to compare those risks.

Risk Categories

There are dozens of risk categories. Topping the list, for purposes of this book, is *environmental risk*. This class of risk is addressed in the last two sections of this chapter. Two other important risks include financial risk and sports risk. Both of these risks are briefly reviewed in the following two sections. Other risks (in alphabetical order) include:

- Aerospace
- Architecture
- Chemical
- Construction
- Education
- Energy
- Governance
- Medical
- Pharmaceutical
- Travel
- Urban Planning

However, irrespective of the risk, it is fair to say that the calculation of risk has now become mandatory in environmental assessment and analysis studies/applications. More and more technical individuals are now required in this field as risk analysis is often performed using the best available data and information.

There are, of course many other risk categories that the engineer, scientist, bureaucrat, society, and so on, are exposed to on a fairly regular basis. Details of these "other" risks are available in the literature. As mentioned earlier, the next two sections explain financial and sports[2] risk. The former topic is directly relevant to environmental risk and the latter is of interest to the author.

The remainder of this chapter addresses the following six topics:

1. Financial Risk
2. Sports Risk
3. Environmental Risk Terms
4. Definitions of Risk-Related Terms
5. Risk Errors
6. Illustrative Examples

FINANCIAL RISK

As noted previously, there are other risks—in addition to environmental ones—that the practicing engineer and applied scientist must be proficient in understanding. Perhaps the most important of these is financial risk. And, although this book is primarily concerned with environmental risk, the author would be negligent if the topic of financial risk was not at least qualitatively addressed.

A company or individual hoping to increase profitability must carefully assess a range of investment opportunities and risks, and select the most profitable options from those available. Increasing competitiveness also requires that efforts need to be made to reduce the costs of existing processes. In order to accomplish this, engineers and scientists should be fully aware of not only technical factors but also economic factors, particularly those that have the largest effect on financial risk and the accompanying topic of profitability.

In earlier years, engineers and scientists concentrated on the technical side of projects and left the financial studies to the economists. In effect, those involved in making estimates of the capital and operating costs have often left the overall economic analysis and investment decision making to others. This approach is no longer acceptable.

Some technical personnel are not equipped to perform a financial or economic analysis. Furthermore, many already working for companies have never taken courses in this area. This short-sighted attitude is surprising for a group of individuals who normally go to great lengths to obtain all the available technical data they can prior to making an assessment of a project or study. The attitude is even more surprising when one notes that data are readily available to enable an engineer or scientist to assess the economic prospects of both his or her own company and those of his or her particular industry.[3]

The term *economic analysis* in real-world problems generally refers to calculations made to determine the conditions for realizing maximum financial return for a design or operation. The same general principles apply whether one is interested in the choice of alternatives for competing projects, in the design of plants so that the various components are economically integrated, or in the economical operation of existing plants. General considerations that form the framework on which sound decisions must be made are often simple. Sometimes their application to the problems encountered in the development of a commercial enterprise involves too many intangibles to allow exact analysis; in that case, judgment must be intuitive. Occasionally, such calculations may be made with a considerable degree of exactness.

Concern with maximum financial return implies that the criterion for judging projects involved is risk and profit. While this is usually true, there are many important objectives, which, though ultimately aimed at increasing profit, cannot be immediately evaluated in quantitative terms. Perhaps the most significant of these is the recent increased concern with environmental degradation, safety, and sustainability. Thus, there has been some tendency in recent years to regard management of commercial organizations as a profession with social obligations and responsibilities; considerations other than the profit motive may govern business decisions. However, these additional social objectives are, for the most part, often not inconsistent with the economic goal of satisfying human wants with the minimum risk. In fact, even in the operation of primarily nonprofit organizations, it is still important to determine the effect of various policies on both risk and long-term economic viability.[4]

If all industrial financial studies simply involved running costs, where a day-to-day expenditure of appropriate raw materials and labor would produce a product of immediate market value, risk predictions as to future demand and prices would be minimized. However, any future return over a period of time can best be evaluated by a host of different methods.

In order to accomplish this, appropriate data for the value of money, i (interest rate), and the lifetime, n, of the process are needed. To a certain extent, the values chosen are interdependent. A large n and a small i can give the same result as a small n and a large i. The result is often evaluated in terms of a lump sum expressed as a present worth factor.

The question now arises as to how the element of risk or uncertainty enters into these formulations. Several characteristics of financial and business risks are of interest in connection with any attempt to formulate methods for taking them into consideration.

First, financial risks are not governed purely by chance like the roll of the dice. What appears as a sound investment to engineers, scientists, and business executives familiar with the know-how and experience in a given company, might recommend a highly speculative venture for a concern engaged largely in a different type of business. Similarly, one investor in the common stock of a given company may not agree with another who does not see growth possibilities in the same stock. Thus, the situation exists where some ventures (investments) require a higher rate of return than others simply because such a rate is necessary to attract venture capital.

Second, aside from the chance of success or failure, a given company is limited in the amount of funds it can invest either from surplus or by borrowing. Thus, in offshore crude oil exploration, a large company (such as British Petroleum) that can finance the drilling of a number of oil wells can recover the costs of unsuccessful ventures from the profits of successful ones and, on the average, show attractive returns, even though four out of five wells drilled turn out to be "dry" holes. The position of the wildcatter or small operator is different in that an unlucky run of failures can put him or her out of business. Companies, like individuals, are limited in the absolute amount of capital they can afford to invest, and, as proposed ventures approach this limit, the rate of return required will increase, even though the financial risk remains unchanged. The utility of a large gain must therefore be balanced against the dis-utility of a smaller loss, which may mean disaster.

In modern business, which is often run by corporations, the entrepreneur is, for the most part, the common stockholder. It is true that the actual operations of the company are in the hands of business executives, and their salaries depend in large part on their ability to show profits. Often, however, their fortunes may not be intimately linked with those of the companies they manage if they own only modest amounts of stock in their companies. Furthermore, their salaries, as reported in the media in recent years—though astronomically high—do not represent a major expense in company operations. If the company they represent fails, they are often able to find opportunities for employment elsewhere. Similarly, the bondholder and preferred stockholder are protected to varying degrees from the risk of company failure. The holder of common stock, on the other hand, is subject to all the risks inherent in running the business. A proper procedure for evaluating new venture capital risk should, therefore, take these factors into consideration.

SPORTS RISK

Another important risk consideration that has become an integral part of modern-day society is sports injury and its associated risks. What sports? Skiing, tennis, football, baseball, boxing, and so on, and, of course, basketball.

Injuries, like accidents, inevitably happen. It is just a matter of when, where, and severity.[2] For athletes, and in particular basketball (a topic that is dear to the author) athletes, injuries can occur at any time during the lifetime of an athlete's participation period; they, for the most part, do not respect age. "Youth" are physically stronger but usually more reckless, while "seniors" are weaker but usually less reckless.[2]

Onto the author's favorite sport. Most basketball injuries involve the knee, either through twisting or through the application of a lateral force. Surgery for such injuries has become much simpler with the invention of the arthroscope. Another common problem is stress fracture: a weakening of the front of the shinbone from overuse, with pain and possible bone cracking as the result. Ligament tears are more common. Almost all these conditions heal with rest. Prevention of injuries depends primarily on good conditioning. The improper or illegal use of drugs and enhancing substances such as steroids for the temporary improvement of athletic performance in all competitions has been a frequent subject of inquiry, but does not fall within the scope of this book.

Unfortunately, risk and risk factors associated with nearly every sport have received only superficial treatment in the literature. Data on comparative risks with various sports are available. Data on comparative sports risks with non-sports activities, for example, driving a car, BBQing, and so on, are also available. But little is available on predicting risks from the probability of occurrence to the consequences associated with the injury. Hopefully this will change in the future.

Most readers are not familiar with health risk concerns with professional football. However, there is concern elsewhere. For example, a May 12, 2014 *New York Times* front page article[5] was titled "Football Risks Sink In, Even In the Heart of Texas: Town Cuts 7th-Grade Tackle Program." The article was about Marshall, Texas where there has been a shift in perceptions about football that would have been hard to imagine when the school made a cameo in the book *Friday Night Lights* nearly 25 years ago. Amid widespread and growing concerns about the physical dangers of the sport, the school board in Marshall approved plans in February of 2014 to shut down the district's entry-level, tackle-football program for seventh graders in favor of flag football. Interestingly, there was little objection. Only time will tell if the decision is the beginning of the end of scholastic football in Texas.

ENVIRONMENTAL RISK TERMS

Is environmental risk important to the practitioner? The reader can decide since all actions, objects, processes, gambling, and so on, entail risk. It is no wonder that this four letter word has become a hot ticket for practitioners.

The next section contains a host of environmental terms and their accompanying definitions. Several of the terms contain the word *risk*. The so-called traditional definitions associated with these words or phrases are presented there. This section attempts to review not only the myriad of risk and risk-related terms but also some of the myriad of accompanying definitions of these terms that have been used in industry and have appeared in the literature. No attempt has been made to present this list in alphabetical order. Rather, this approach has attempted to provide the various terms in a logical, sequential order.

The four major risk terms include

1. Risk
2. Risk assessment
3. Risk analysis
4. Risk management

"Risk" was defined earlier as a measure of economic loss, human injury, or human health effect, and in terms of both the likelihood (probability) and the magnitude (consequences) associated with either a loss or injury. Although it is a quantifiable term, it has been misused by practitioners. "Risk assessment" involves the process of determining the events or problems that can produce a risk, the corresponding probabilities and consequences, and finally, the characterization of the risk. "Risk analysis" employs the results of the aforementioned risk assessment and attempts to optimally use these results; in effect, it analyzes risk assessment information. Finally, "risk management" uses all the information provided by the risk assessment and risk analysis steps to reduce or eliminate the risk, select the optimum action(s), or evaluate the net benefits verses health/safety concerns. Note, however, that in line with its title, this book is primarily concerned with risk calculations from both a health and hazard perspective.

Of course, there are other risks. The definitions (for purposes of this section) for these other risks are as follows: "Individual risk" is defined as the risk to an individual; this can include a health problem or injury, the likelihood of occurrence, and the time period over which the problem might occur. The "maximum individual risk" is the aforementioned individual risk to a person exposed to the highest risk in an exposed population; this can be determined by calculating individual risks at every "location" and selecting the result for the maximum value. The "average individual risk" (in an *exposed* population) is the aforementioned individual risk averaged over the total population that is exposed to the risk in question. Alternatively, the average individual risk (in a *total* population) is the individual risk averaged over the entire population without regard to whether or not all the individuals in the population are actually exposed to the risk. Unfortunately, this particular average risk—whether applied to employees or the public—can be (at times) extremely misleading. These average risks have, on occasion, been expressed as exposed hours per worked hours; thus, the risk may be calculated for a given duration of time or averaged over the working day. "Societal risk" provides a measure of risk to a specific group of people, that is, it is based on the people affected by an event or scenario. "Time to respond risk" characterizes the time that a response occurs following a given event/scenario. "Risk communication" is concerned with communicating the information generated from a risk assessment, risk analyses, and a risk management study. "Ecological risk" is a risk that describes the likelihood that adverse ecological effects resulting from an event/scenario will occur. "Total risk" is the term generally employed to describe the summation of the risk from all scenarios. "De minimus risk" has recently taken on significant importance in toxicology. It is defined as a risk judged to be too insignificant to be of societal concern or too small to be effectively applied to standard risk assessment studies. Financial risk important enough to receive treatment in an earlier section. Finally, the new kid on the block is "unreasonable risk," a term that has yet to be clearly defined. Perhaps it would be best to simply say that unreasonable risk is unreasonable in comparison to some other risk that is reasonable, that is, when one compares two risks and selects one as more reasonable.

DEFINITIONS OF RISK-RELATED TERMS

This section defines many of the terms that the reader will encounter in this book. The following list is therefore not a complete glossary of all the terms that appear in the risk and risk-related fields. It should also be noted that many of the terms have come to mean different things to different people; this will become evident as one delves deeper into the literature.[1]

Acute (risk): Risk associated with short periods of time. For health risk, it usually represents short exposures to high concentrations of a hazardous agent.

Auto-ignition temperature: The lowest temperature at which a flammable gas in air will ignite without an ignition source.

Average rate of death (ROD): The average number of fatalities that can be expected per unit time (usually on an annual basis) from all possible risks and/or incidents.

C (ceiling): The term used to describe the maximum allowable exposure concentration of a hazardous agent related to industrial exposures to hazardous vapors.

Cancer: A tumor formed by mutated cells.

Carcinogen: A cancer-causing chemical.

CAS (Chemical Abstract Service): CAS numbers are used to identify chemicals and mixtures of chemicals.

Catastrophe: A major loss in terms of death, injuries, and damage.

Cause–consequence: A method for determining the possible consequences or outcomes arising from a logical combination of input events or conditions that determine a cause.

Chronic (risk): Risks associated with long-term chemical exposure, usually at low concentrations.

Conditional probability: The probability of occurrence of an event given that a precursor event has occurred.

Confidence interval: A range of values of a variable with a specific probability that the true value of the variable lies within this range. The conventional confidence interval probability is the 95% confidence interval, defining the range of a variable in which its true values falls with 95% confidence.[6]

Confidence limits: The upper and lower range of values of a variable defining its specific confidence interval.

Consequences: A measure of the expected effects of an incident outcome or cause.

CPQRA (chemical process quantitative risk analysis): It is analogous to a hazard risk assessment (HZRA).

Deflagration: The chemical reaction of a substance in which the reaction front advances into the unreacted substance present at less than sonic velocity.

Delphi method: A polling of experts that involves the following:

1. Select a group of experts (usually three or more).
2. Solicit, in isolation, their independent estimates of the value of a particular parameter and their reason for the choice.
3. Provide initial analysis results to all experts and allow them to then revise their initial values.
4. Use the average of the final estimates as the best estimate of the parameter.[6] Use the standard deviation of the estimates as a measure of uncertainty.[6]

The procedure is iterative, with feedback between iterations. The author modestly refers to it at the "Theodore method."[7]

Dermal: Applied to the skin.

Detonation: A release of energy caused by a rapid chemical reaction of a substance in which the reaction front advances into the unreacted substance present at greater than sonic velocity.

Domino effects: The triggering of secondary events; usually considered when a significant escalation of the original incident could result.

Dose: A weight (or volume) of a chemical agent, usually normalized to a unit of body weight.

Episodic release: A massive release of limited or short duration, usually associated with an accident.

Equipment reliability: The probability that, when operating under stated conditions, the equipment will perform its intended purpose for a specified period of time.

Ergonomics: The interaction between humans and their environment, usually in an industrial or other man-made setting.

Event: An occurrence associated with an incident either as the cause or a contributing cause of the incident, or as a response to an initiating event.

Event sequence: A specific sequence of events composed of initiating events and intermediate events that may lead to a problem or an incident.

Event tree analysis (*ETA*): A graphical logic model that identifies and attempts to quantify possible outcomes following an initiating event.

Explosion: A release of energy that causes a pressure discontinuity or blast wave.

Exposure period: The duration of an exposure.

External event: A natural or man-made event; often an accident.

Failure frequency: The frequency (relative to time) of failure.

Failure mode: A symptom, condition, or manner in which a failure occurs.

Failure probability: The probability that failure will occur, usually for a given time interval.

Failure rate: The number of failures divided by the total elapsed time during which these failures occur.

Fatal accident rate (FAR): The estimated number of fatalities per 10^8 exposure hours (roughly 1000 employee working lifetimes).

Fault tree: A method of representing the logical combinations of events that lead to a particular outcome (top event).

Fault tree analysis (FTA): A logic model that identifies and attempts to quantify possible causes of an event.

Federal Register: A daily government publication of laws and regulations promulgated by the U.S. Federal Government.

Flammability limits: The range in which a gaseous compound in air will explode or burst into flames if ignited.

Frequency: Number of occurrences of an event per unit time.

Half-life: The time required for a chemical concentration or quantity to decrease by half its current value.

Hazard (problem): An event associated with an accident which has the potential for causing damage to people, property, or the environment.

Hazard and operability study (HAZOP): A technique to identify process hazards and potential operating problems using a series of guide words that key-in on process deviation.

Hazard risk assessment (HZRA): A technique associated with quantifying the risk of a hazard problem employing probability and consequence information.

Health (problem): A problem normally associated with health arising from the continuous emission of a chemical into the environment.

Health risk assessment (HRA): A technique associated with quantifying the risk of a health problem employing toxicology and exposure information.

Human error: Actions by engineers, operators, managers, and so on, that may contribute to or result in accidents.

Human error probability: The ratio between the number of human errors and the number of opportunities for human error.

Human factors: Factors attempting to match human capacities and limitations.

Human reliability: A measure of human errors.

Incident: An event.

Individual risk: The risk to an individual.

Ingestion: The intake of a chemical through the mouth.

Initiating event: The first event in an event sequence.

Instantaneous release: Emissions that occur over a very short duration.

Intermediate event: An event that propagates or mitigates the initiating event during an event sequence.

Isopleth: A concentration plot at specific locations, usually downwind from a release source.

Lethal concentration (LC): The concentration of a chemical that will kill a test animal, usually based on a 1–4 h exposure duration.

Lethal concentration (LC_{50}): The concentration of a chemical that will kill 50% of test animals, usually based on a 1–4 h exposure duration.

Lethal dose (LD): The quantity of a chemical that will kill a test animal, usually normalized to a unit of body weight.

Lethal dose 50 (LD_{50}): The quantity of a chemical that will kill 50% of test animals, usually normalized to a unit of body weight.

LEL/LFL: The lower explosive/flammability limit of a chemical in air that will produce an explosion or flame if ignited.

Level of concern (LOC): The concentration of a chemical above which there may be adverse human health effects.

Likelihood: A measure of the expected probability or frequency of occurrence of an event.

Malignant tumor: A cancerous tumor.

Mutagen: A chemical capable of changing a living cell.

PEL (permissible exposure limit): The permissible exposure limit of a chemical in air, established by the Occupation Safety and Health Administration (OSHA).

Personal protection equipment (PPE): Material/equipment worn to protect a worker from exposure to hazardous agents.

Precision: The degree of "exactness" of repeated measures relative to the true or actual value.

ppm: The parts per million of a chemical in air—almost always on a volume basis; often designated as ppmv as opposed to ppmm (mass basis).

ppb: The part per billion of a chemical in air—almost always on a volume basis; often designated as ppbv as opposed to ppbm (mass basis).

Maximum individual risk: The highest individual risk in an exposed population.

Probability: An expression for the likelihood of occurrence of an event or an event sequence, usually over an interval of time.

Propagating factors: Influences that contribute to the sequence of events following the initiating event.

Protective system: Systems, such as pressure vessel relief valves, that function to prevent or mitigate the occurrence of an accident or incident.

Risk: A measure of economic loss or human injury in terms of both the incident likelihood and the magnitude of the loss or injury.

Risk analysis: The engineering evaluation of incident consequences, frequencies, and risk assessment results.

Risk assessment: The process by which risk estimates are made.

Risk contour: Lines on a risk graph that connect points of equal risk.

Risk estimation: A combination of the estimated consequences and likelihood of a risk.

Risk management: The application of management policies, procedures, and practices in analyzing, assessing, and controlling risk.

Risk perception: The perception of risk that is a function of age, race, sex, personal history and background, familiarity with the potential risk, dread factors, perceived benefits of the risk causing action, marital status, residence, and so on.

Societal risk: A measure of risk to a group of individuals.

Source term: The estimation of the release of a hazardous agent from a source.

Time of failure: The time period associated with the inability to perform a duty or intended function.

TLV (threshold limit value): Established by the American Council of Governmental Industrial Hygienists, it is the concentration of a chemical in air that produces no adverse effects.

TLV-TWA: The allowable time weighted average concentration of a chemical in air for an 8 h workday/40 h workweek that produces no adverse effect.

TLV-STEL: The short-term exposure limit (maximum concentration in air) for a continuous 15 min averaged exposure duration.

TLV-C: The ceiling exposure limit representing the maximum concentration of a chemical in air that should never be exceeded.

Toxic dose: The combination of concentration and exposure period for a toxic agent to produce a specific harmful effect.

UEL/UFL: The upper explosive/flammability limit of a chemical in air that will produce an explosion or flame if ignited.

Uncertainty: A measure, often quantitative, of the degree of doubt or lack of certainty associated with an estimate.

Finally, the reader should carefully note the difference between the definitions of *health* and *hazard*.

RISK ERRORS

No discussion of risk would be complete without some mention of *errors, accuracy, and precision*. The significance of conclusions based upon numerical results is necessarily determined by the reliability of the data and of the methods of calculation in which they are employed. It should be understood at the outset that most numerical calculations are by their very nature *inexact*. The errors are primarily due to one of three sources: inaccuracies in the original data, lack of precision in carrying out elementary operations, and inaccuracies introduced by approximate methods of solution.

Of particular significance in some applications are the errors due to *roundoff* and the inability of carrying, in a given calculation, more than a certain number of significant figures. Terms such as *absolute error, relative error,* and *truncation error* have a very real meaning, and an analysis parallel to another in question must frequently be carried out to establish the reliability of a given answer.

Accuracy is a term often employed in any discussion of risk. Accuracy is defined[8] as "the quality of state of being accurate or exact... precision... exactness... free from mistakes or errors...adhering closely to a standard." Precision is defined[8] as "the quality of being precise...strictly defined; accurately stated; definite... with no variations... minutely exact...strictly conforms to usage; scrupulous; fastidious." The *accuracy* is poor if, for example, the reported weight of a beaker is 35 g when the actual weight is 55 g. If the weight is reported as 35.29 g, the reading is more *precise*. Therefore, data or readings or calculations upon which probability distribution may be based are subject to errors... and these errors may also be subject to distributions. The reader should note that this subject area is beyond the scope of this book.

Standard statistical techniques can be utilized to determine an estimate for an error in a specific piece of data where that data point is calculated from several other previously determined pieces of data. The error to be determined is known as the "estimated accumulated error." Unfortunately, most engineering textbooks give the impression that there should be a 100% match between data points and graphs drawn from theoretical equations. As discussed above, real-world data is limited in its accuracy by experimental hardware, measuring devices, and other factors; therefore, it is important that engineers become familiar with one or more methods for determining the magnitude of errors associated with data and calculations.

The subject of error in data recorded and the propagation of individual errors from various other data is a topic that requires careful attention and study. The focus of this presentation is to handle the relatively straightforward topic of the estimated accumulated error for calculations. For any function Y, the estimated accumulated error, S_Y, is defined by Equation 1.1

$$S_Y^2 = \sum_{i=1}^{N} a_i^2 \cdot S_i^2$$

where

$$a_i = \left(\frac{\partial Y}{\partial u_i} \right) \text{ at } u_i = u_i \tag{1.1}$$

and

u_i is the independent variable in the defining equation for Y.
N is the number of independent variables.
S_i is the estimate of the errors for each independent variable u_i.

The reader is referred to the literature for additional details and sample calculations.

What about *uncertainty*? Qualitatively, uncertainty may be viewed as having two components: variability and lack of knowledge. Uncertainty, whether applied to toxicological values, probability, consequences, risks, and so on, may be described qualitatively or quantitatively. Qualitatively, descriptions include large, huge, monstrous, tiny, very small, and so on. Quantitative terms describing the uncertainty associated with a value x are normally in the form $x \pm w_{zx}$, where w_{zx} provides a measure of the uncertainty (e.g., standard deviation, 95% confidence limit, etc.).[5]

A substantial amount of information on uncertainty and uncertainty analysis is available. Useful references abound but, in general, there are three main sources of uncertainty that have been earmarked by practicing engineers and scientists:

1. Model uncertainty
2. Data uncertainty
3. General quality uncertainty

"Model uncertainty" reflects the weaknesses, deficiencies, and inadequacies present in any model and may be viewed as a measure of the displacement of the model from reality. "Data uncertainty" results from incomplete data measurement, estimation, inference, or supposed expert opinion. "General quality uncertainty" arises because the practitioner often cannot identify every health problem or hazard incident. Naturally, the risk engineer's objective is to be certain that the major contributors to the risk are identified, addressed, and quantified. Uncertainty here arises from not knowing the individual risk contributions from those risk problems that have not been considered; one, therefore, may not be able to accurately predict the overall (or combined) risk. Byrd and Cothern[9] have expanded this three-part uncertainty categorization in the following manner:

1. Subjective judgment
2. Linguistic imprecision
3. Statistical variation
4. Sampling error
5. Inherent randomness
6. Mathematical modeling
7. Causality
8. Lack of data or information
9. Problem formulation

Sensitivity and *importance* are also issues in the utilization of risk results. As noted earlier, uncertainty analysis is used to estimate the effect of data and model uncertainties on the risk estimate. Sensitivity analysis estimates the effect on calculated outcomes of varying inputs to the models individually or in combination. Importance analysis quantifies and ranks risk estimate contributions from subsystems or components of the complete system (e.g., individual incidents, groups of incidents, sections of a process, etc.).

To summarize, different assumptions can change any quantitative risk characterization by several orders of magnitude. The uncertainty that arises is related to how well (and often, consistently) input data are obtained, generated, or measured, and the degree to which judgment is involved in developing risk scenarios and selecting input data. Simply put by some, uncertainty arises from how data/evidence was both measured and interpreted. Despite these limitations and uncertainties, risk characterizations provide the practitioner with some analysis and assessment capabilities.

Additional information is provided in the next chapter.

ILLUSTRATIVE EXAMPLES

Illustrative Example 1.1

Provide a qualitative definition of risk.

Solution

Risk can be defined as the product of two factors: (1) the probability of an undesirable event, and (2) the measured consequences of the undesirable event. Measured consequences may be stated in terms of financial loss, injuries, deaths, or other variables.

Illustrative Example 1.2

Define *failure*.

Solution

Failure represents an inability to perform some required function.

Illustrative Example 1.3

Define *reliability*.

Solution

Reliability is the probability that a system or one of its components will perform its intended function under certain conditions for a specified period. The reliability of a system and its probability of failure are complementary in the sense that the sum of these two probabilities is unity. The basic concepts and theorems of probability that find application in the estimation of risk and reliability are considered in the next chapter.

Illustrative Example 1.4

Compare annual versus lifetime risks.

Solution

A time frame must be included with a risk estimate for the numbers to be meaningful. For both health and hazard risks, annual or lifetime risks are commonly used. Direct evidence is usually expressed annually because the information is often collected and summarized annually. However, predictive information is commonly expressed as a lifetime probability, for example, when expressing cancer risk or a terrorist-related risk.

Illustrative Example 1.5

Define an *accident*.

Solution

As noted earlier, an accident is an unexpected event that has undesirable consequences and can be quantitatively described through a HZRA. The causes of accidents have to be identified in order to help prevent accidents from occurring. Any
situation or characteristic of a system, plant, or process that has the potential to
cause damage to life, property, or the environment is considered a *hazard*. A hazard
can also be defined as any characteristic that has the potential to cause an accident.

Illustrative Example 1.6

Describe what a cancer risk number of 10^{-6} probability means.

Solution

A cancer risk number usually represents the probability of developing cancer risk.
A risk of 10^{-6} indicates an individual has a 1 in 1,000,000 chance of developing cancer throughout a lifetime (assumed to be 70 years). One generally can
also assume an upper 95% confidence limit on the maximum likelihood estimate.
Since the predicted risk is an upper bound, the actual risk is unlikely to be higher
but may be lower than the predicted risk.

Illustrative Example 1.7

Qualitatively describe the following terms:

1. Substantial risk
2. Unreasonable risk
3. Insignificant risk
4. De minimus risk
5. Acceptable risk

Refer to the earlier section concerned with definitions of risk-related terms.

Solution

1. Each individual has a different concept of the meaning of the term *substantial* as it applies to health.
2. *Unreasonable* risk is an unacceptable risk, a risk level that is significantly
   (usually 1–2 orders of magnitude) above the EPA quantitative risk standard for lifetime probabilities of adverse effects.
3. The EPA has generally described *insignificant* risk quantitatively in terms
   of lifetime probabilities below which the risk is assumed low enough to
   be ignored.
4. *De minimus* risk is analogous to insignificant risk and has become a legal
   term that is decided on a case-by-case basis.
5. *Acceptable* risk generally suggests that the risk is either insignificant or
   perhaps zero.

REFERENCES

1. L. Theodore and R. Dupont, *Environmental Health Risk and Hazard Risk Assessment: Principles and Calculations*, CRC Press/Taylor & Francis Group, Boca Raton, FL, 2013.
2. L. Theodore, *Basketball Coaching 101*, Theodore Tutorials, East Williston, NY, 2015.
3. L. Theodore and F. Ricci, *Mass Transfer Operations for the Practicing Engineer* (adapted from), John Wiley & Sons, Hoboken, NJ, 2010.
4. J. Reynolds, J. Jeris, and L. Theodore, *Handbook of Chemical and Environmental Engineering Calculations*, John Wiley & Sons, Hoboken, NJ, 2004.
5. *New York Times*, Football Risks Sink In, Even in the Heart of Texas, New York, NY, May 12, 2014.
6. S. Shaefer and L. Theodore, *Probability and Statistics Applications for Environmental Science* (adapted from), CRC Press, Boca Raton, FL, 2007.
7. L. Theodore, personal notes, East Williston, NY, 1985.
8. *Webster's College Dictionary*, 2nd edn., Prentice Hall, Upper Saddle River, NJ, 1971.
9. D. Byrd and C. Cotherm, *Introduction to Risk Analysis*, Government Institutes, Rockville, MD, 2000.

2 Environmental Risk Analysis

INTRODUCTION

People face all kinds of risks every day, some voluntarily and others involuntarily. Therefore, risk plays a very important role in today's world. Studies on cancer caused a turning point in the world of risk because it opened the eyes of risk scientists and health professionals to the world of health risk assessments (HRAs).

The usual objective of HRA and the accompanying calculations is to evaluate the potential for adverse health effects from the release of chemicals into the environment. Unfortunately, the environment is very complex since there is a vast array of potential receptors present. The task of testing and evaluating each of the myriad number of chemicals on the market for their impact on human populations and ecosystems becomes extremely difficult. To further complicate the problem, health is a concept that has come to mean different things to different people. Some have defined it as follows: "… a state of complete physical, mental and social well-being and not merely the absence of disease or infirmary." Many other definitions and concepts have been purposed and appear in the literature.

Since 1970, the field of HRA has received widespread attention within both the scientific and the regulatory communities. It has also attracted the attention of the public. Properly conducted risk analyses and risk assessment calculations have received fairly broad acceptance, in part because they put into perspective the terms toxic, health, and risk. Toxicity plays a role and is an inherent property of all substances. It states that all chemical and physical agents can produce adverse health effects at some dose or under some specific exposure conditions.

Risk assessment of accidents serves a dual purpose. It estimates the probability that an accident will occur and also assesses the severity of the consequences of an accident. Consequences may include damage to the surrounding environment, financial loss, injury, or loss of life. This chapter is also concerned with introducing the reader to the methods used to identify these hazards and the causes and consequences of accidents. Risk assessment of accidents (or hazard risk assessment [HZRA]) provides an effective way to help ensure that a mishap either does not occur or reduces the likelihood of severe consequences as a result of the accident. The results of the risk assessment allow concerned parties to take precautions to prevent an accident before it happens.

There are other classes of environmental health risks that do not pertain to chemicals but are an integral part of HZRA. For example, health problems can arise immediately/soon after a hazard, such as a hurricane or earthquake, that can leave local inhabitants without potable water for an extended period of time.

Environmental risk assessment may be broadly defined as a scientific enterprise in which facts and assumptions are used to estimate the potential for adverse effects. Risk management, as the term is used by the U.S. Environmental Protection Agency (EPA) and other regulatory agencies, refers to a decision-making process which involves such considerations as risk assessment, technological feasibility, economic information about costs and benefits, statutory requirements, public concerns, and other factors. Risk communication is the exchange of information about risk.

Risk assessment may also be defined as the characterization of potential adverse effects to humans or to an ecosystem resulting from environmental hazards. Risk assessment supports risk management, the set of choices centering on whether and how much to control future exposure to the suspected hazards. Risk managers face the necessity of making difficult decisions involving uncertain science, potentially grave consequences to health or the environment, and large economic effects on industry and consumers. What risk assessment provides is an orderly, explicit, and consistent way to deal with scientific issues in evaluating whether a health problem or a hazard exists. This evaluation typically involves large *uncertainties*, because the available scientific data are limited, and the mechanisms for adverse health impacts or environmental damage are only imperfectly understood.

From a risk management standpoint, whether dealing with a site-specific situation or a national standard, the deciding question is ultimately, "What degree of risk is acceptable?" In general, this does not mean a "zero risk" standard, but rather a concept of *negligible* risk. At what point is there really no significant health or environmental risk, and at what point is there an adequate safety margin to protect public health and the environment? In addition, some environmental statutes require consideration of benefits together with risks in making risk management decisions.

Thus, it should be noted that health risk addresses risks that arise from health and health-related problems. Chemicals are generally the culprit. Both the effect on and exposure to a receptor (in this case, generally a human) ultimately determine the risk to the individual for the health problem of concern. The risk can be described in either qualitative or quantitative terms, and there are various terms that may be used, for example, 10 individuals will become sick, or 1×10^{-6} (one in a million) will die, or something as simple as "it is a major problem."

Another category of environmental risk is the aforementioned hazard risk. This class of risk is employed to describe risks associated with hazards or hazard-related problems, for example, accidents, negative events, and catastrophes. Unlike most health problems, these usually occur over a short period of time, say a few seconds or minutes. Both the probability and the consequence associated with the accident/event ultimately determine the hazard risk. Once again, the risk can be described in either qualitative or quantitative terms, and there are various terms that may be used.

As noted earlier, once a risk has been calculated, one needs to gauge the estimated consequences (or opportunities if examining financial/economic scenarios), and evaluate and prioritize options for risk management or mitigation. These potentially strategic evaluations are usually fraught with uncertainties at numerous levels. Thus, the risk assessment process is normally followed by option analyses; these options are usually based on decision-making procedures that are beyond the scope of this book. However, it is fair to say that there may be a full range of outcomes and consequences to various scenarios. It should also be noted that risk assessment is a dynamic process that can very definitely be a function of time. Much of this material is addressed later in the chapter.

The remaining sections of this chapter address the following four topics.

1. Health Risk Assessment/Analysis
2. Hazard Risk Assessment/Analysis
3. Risk Uncertainties/Limitations
4. Illustrative Examples

Note that the bulk of the material for this chapter has been adopted from the literature.[1–3]

HEALTH RISK ASSESSMENT/ANALYSIS

Health risk assessments (HRAs) provide an orderly, explicit, and consistent way to deal with issues in evaluating whether a health problem exists and what the magnitude of the problem may be. This evaluation typically involves large uncertainties (to be discussed in a later section) because the available scientific data are limited and the mechanisms for adverse health impacts or environmental damage are only imperfectly understood.

When one examines risk, how does one decide how safe is "safe," or how clean is "clean?" To begin with, one has to look at both inputs of the risk equations, that is, both the toxicity of a pollutant and the extent of exposure. Information is required for both the current and the potential exposure, considering all possible exposure pathways. In addition to human health risks, one needs to look at the potential ecological or other environmental effects.

In recent years, several guidelines and handbooks have been published to help explain approaches for conducting HRAs. As discussed by a special National Academy of Sciences Committee which convened in 1983, most human or environmental health hazards can be evaluated by dividing the analysis into four parts: Health problem identification, dose–response assessment or toxicity assessment, exposure assessment, and risk characterization (see Figure 2.1). The risk assessment might stop with the first step, health problem identification, if no adverse effect is identified or if an agency elects to take regulatory actions without further analysis.[4] Regarding identification, a health problem is defined as a toxic agent or a set of conditions that has the potential to cause adverse effects to human health or the environment. Health problem identification involves an evaluation of various forms of information in order to identify the different problems. Dose–response or toxicity assessment is also required in an overall assessment; responses and effects can vary widely since all chemicals and contaminants vary in their capacity to cause adverse effects. This step frequently requires that assumptions be made to relate experimental results from animal tests to expected effects on exposed humans. Exposure assessment is the determination of the magnitude, frequency, duration, and routes of exposure of toxic agents to human populations and ecosystems. Finally, in health risk characterization,

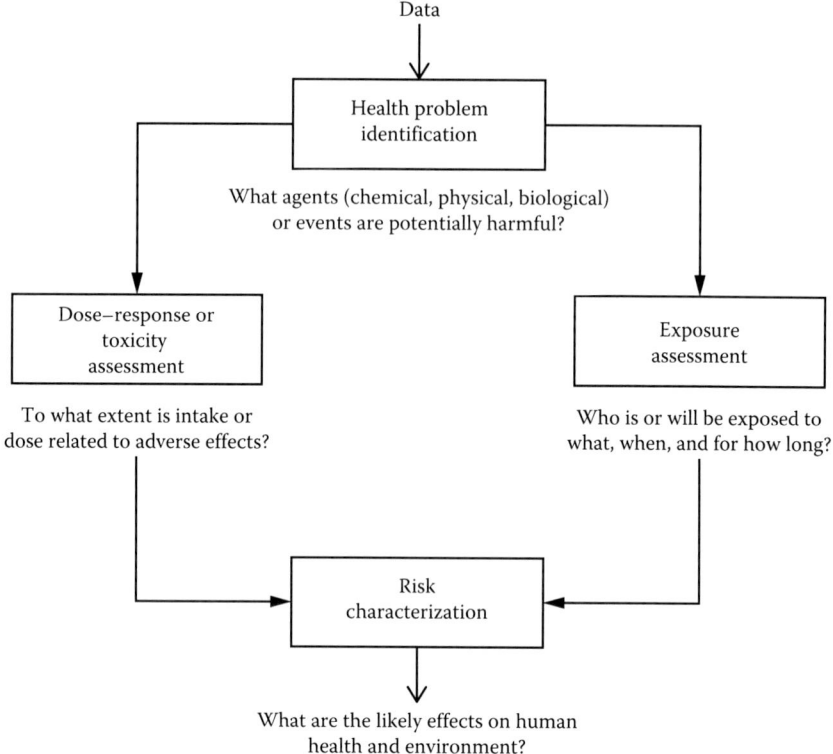

FIGURE 2.1   Health risk evaluation process.

toxicology and exposure data/information are combined to obtain a qualitative or quantitative expression of risk. An expanded presentation on each of the four HRA steps is provided in the next four sections.

Health Problem Identification

Health problem identification is defined as the process of determining whether human exposure to a chemical at some dose could cause an increase in the incidence of an adverse health condition (cancer, birth defect, etc.), or whether exposure to nonhumans, such as fish, birds, and other forms of wildlife, could cause adverse effects. In other words, does exposure to a chemical have the potential to cause harm? It involves characterizing the quality, nature, and strength of the evidence of causation. It may not give a yes or no answer; however, it is intended to provide an assessment on which to base a decision as to whether a health problem has been identified. This identification characterizes the problem in terms of the agent and dose of the agent. Since there are few hazardous chemicals or hazardous agents for which definitive exposure data in humans exist, the identification of health hazards is often characterized by the effects of health problems on laboratory test animals or other species and test systems.[5]

There are numerous methods available to identify the potential for chemicals to cause both adverse health conditions and significant effects on the environment. These can include, but are not limited to toxicology, epidemiology, molecular and atomic structural analysis, material safety data sheets (MSDSs), standardized mortality ratios (observed deaths or expected deaths), engineering approaches to problem solving, analysis of the fate of chemicals in the environment, and evaluations of carcinogenic versus noncarcinogenic health hazards.

TOXICITY AND DOSE–RESPONSE

Dose–response assessment is the process of characterizing the relationship between the dose of an agent administered or received and the incidence of an adverse health effect in exposed populations. This process considers such important factors as intensity of exposure, age, pattern of exposure, and other variables that might modify the response, such as sex and lifestyle. In effect, it involves the evaluation of the effects expected from various quantity/concentration levels of a particular chemical in the environment. Dose and response are therefore fundamental concepts that provide a relationship between the dosage of a toxic agent and the biological response. The magnitude of the biological response depends on the concentration of the contaminant/physical agent at the site of action, while the concentration of the contaminant at the active site depends on the dose. Thus, the dose and the response are causally related. Toxicity data exhibit a dose–response *relationship* if a mathematical model can be formulated to describe the response of the receptor and/or test organism in terms of the dose administered. The relation often takes the form of a percentage or number of receptors responding in a given manner either to a dose or to a specified range of concentrations over a given period of time. A dose–response assessment usually requires extrapolation from high to low doses and from animal to humans or one laboratory animal species to a wildlife species. A dose–response assessment should also describe and justify the methods of extrapolation used to predict incidence, and it should characterize the statistical and biological uncertainties in these methods. When possible, the uncertainties should be described numerically rather than quantitatively (see the section "Risk Uncertainties/Limitations" later in this chapter).

Why is toxicology so important? As noted earlier, it is the dose that makes the poison. A low-level dose may cause no effect. Yet, a larger dose may lead to either an adverse health effect or even death. This dose variation is also a function of the chemical of concern. Furthermore, the manner in which the dose impacts a chemical's absorptions, distribution, metabolism in the human body, and ultimate excretion from the body can vary with both the chemical and the dose.

Once again, it is important to differentiate between the terms *chronic* and *acute* as they relate to toxicity. *Chronic toxicity* is caused by long-term or repeated exposure to low doses of the chemical, and the intensity is usually less than with acute exposures. *Acute toxicity* is caused by large doses of a chemical over short time periods, as when the time between the exposure and the onset of a health problem is short. The intensity of acute toxicity effects is usually greater than that of chronic toxicity exposure, since the receptor(s) usually have little opportunity to detoxify, eliminate, or adapt to the administered dose of an acutely toxic agent.

The classification as to whether a chemical agent is a carcinogen or a noncarcinogen can help identify whether it is a health problem. Toxicity, that is, the degree to which a chemical is considered a health problem, is characterized by its threshold. A threshold or lower limit below which effects cannot be observed characterizes the dose–effect or dose–response relationship. These effects can occur at the cellular, subcellular, and/or molecular level. The body protects itself against toxic chemicals with repair mechanisms and by attributing critical functions to large numbers of the same units. Thus, in order for a toxic effect to occur, a number of these units greater than the threshold for the target dose must be affected. For example, carbon tetrachloride is a solvent that causes disease in the liver tissue. The body's repair mechanism allows it to replace lost cells with new cells, thus allowing the liver to continue to function. Beyond a certain threshold, however, the liver cannot function and the damage cannot be reversed.

Noncarcinogenic effects include all toxicological responses except tumors. Toxicological responses and mechanisms vary widely, and examples of these include interference with normal cell processes by displacing elements out of the cell and binding with a cell to reduce membrane permeability. However, the majority of noncarcinogenic effects involve enzymes. In the body, different enzymes perform specific functions. When an enzyme binds with a toxic substance, the enzyme may be prohibited from performing its function properly, thereby exhibiting a toxic response.

Carcinogens cause cancer. Cancerous cells are normal cells that become abnormally altered and divide uncontrollably. The disease of cancer is characterized by tumors or neoplasms (meaning "new growth"); however, not all tumors are cancerous. Benign tumors are not cancerous and do not spread. Malignant tumors are cancerous and spread, or metastasize, to surrounding "structures." This invasion of surrounding structures by malignant tumors occurs because the abnormal alteration of the cells prevents them from responding to the body's regulatory signals that control cell growth.

Carcinogenesis is the process that occurs during exposure to a carcinogenic chemical and the development of malignancy. The three stages of carcinogenesis are *initiation*, *promotion*, and *progression*. Initiation is the alteration or mutation of a normal cell into an abnormal cancerous cell. Promotion is the increase in the replication rate and number of initiated cells. It is caused by promoting carcinogens. The promoter is not usually the same carcinogen that initiated the first stage of carcinogenesis. All cells that have been initiated and promoted do not develop into malignant cells. The body's defense mechanism against foreign substances—the immune system—recognizes and rejects some of these cells. In progression, the third stage of carcinogenesis, the abnormal cells invade surrounding tissues and spread to distant organ sites. Progression involves more genetic mutations than those required in initiation and promotion.

EXPOSURE ASSESSMENT

As noted, a critical component of environmental health risk assessment (HRA) is exposure assessment. It is defined as the determination of the concentration of chemicals in time and space at the location of receptors and/or target populations. This description must therefore also include an identification of all major pathways for movement and transformation of a toxic material from a source to receptors. Ideally, concentrations should be identified as a function of time and location, and should include all major transformation processes. The principal pathways generally considered in exposure assessments are atmospheric transport and surface and groundwater transport. Since atmospheric dispersion has received the bulk of treatment in the literature, a good part of the material to follow will address this topic.

The exposure assessment process consists of two basic methods for determining the concentration of a chemical to which receptor target populations are exposed:

1. The first is the direct measurement of the intensity, frequency, and duration of human or animal exposure to a pollutant currently present in the environment. This is a common practice in occupational settings.
2. In some situations, however, either concentrations are too low to be detected against the background, or direct measurement is too costly or difficult to implement. Under these circumstances, the second method is employed. It involves the use of mathematical models to estimate hypothetical exposures that might arise from the release of new chemicals into the environment. This section discusses some of these models.

In its most complete form, an exposure assessment should describe the magnitude, duration, timing, and route of exposure of the hazardous agent, along with the size, nature, and classes of the human, animal, aquatic, or wildlife populations exposed, and the uncertainties in all estimates. The exposure assessment can often be used to identify feasible prospective control options and predict the effects of available treatment technologies for controlling or limiting exposure.[5] However, the estimation of the likelihood of exposure to a chemical remains a difficult task. Attention in the past focused on too many overly conservative assumptions. This, in turn, resulted in an overestimation of the actual exposure risk posed to vulnerable receptors.

Obviously, without exposure(s), there are no risks. To experience adverse effects, one must first come into contact with the toxic agent(s). Exposures to chemicals can occur via inhalation of air (breathing), intake into the body via ingestion of water and food (drinking and eating), or adsorption through the skin. These intake processes are followed by chemical distribution through the body via the bloodstream. After being absorbed and distributed, the chemical(s) may be metabolized and excreted, either as the parent compound or as their metabolites or their conjugates adduct. The principal excretory organs are the kidney, liver, and lungs.

As noted earlier, the main pathways of exposure considered in human exposure assessments are via atmospheric, surface, and groundwater transport. However, the ingestion of toxic materials that have passed through the aquatic and terrestrial food chains, and dermal absorption, are two other pathways of potentially significant human exposure.

The physical and chemical properties of the chemical under study will dictate the primary route(s) by which exposure will occur. Naturally, the chemical under study should be analyzed for the primary route(s) of human exposure. There are instances where humans may be exposed to a compound by more than one route, for example, by inhalation and oral ingestion. Which is the most significant route of administration? Assuming approximately equal exposure by both routes, it is recommended that the chemical exposure assessment should focus on the route posing the greater risk. For those situations where one route of exposure predominates over another, the dominant route should be considered. Once an exposure assessment determines the quantity of a chemical with which human populations may come in contact, the information can be combined with toxicity data to estimate potential health risks.[1]

The reader should once again note that two general types of potential health risk from chemical exposures exist. These are classified as follows:

1. *Chronic*: Risk related to continuous exposures over long periods of time, generally several months to a year. Concentrations of emitted chemicals are usually relatively low. This subject area falls in the general domain of HRA, and it is this subject that is addressed in this section. Thus, in contrast to the acute (short-term) exposures that predominate in hazard risk assessments (HZRAs), chronic (long-term) exposures are the major concern in HRAs.

2. *Acute*: Risk related to exposures that occur for a relatively short period of time, generally from minutes to 1–2 days. Concentrations of emitted chemicals are usually high relative to their no-effect levels. In addition to inhalation, airborne substances might directly contact the skin, or liquids and sludges may be splashed on the skin or into the eyes, leading to adverse health effects in acute risk settings. This subject area falls, in a general sense, in the domain of HZRA and is addressed in the next section.

HEALTH RISK CHARACTERIZATION

Health risk characterization is the process of estimating the incidence of a health effect under the various conditions of human or animal exposure described in an exposure assessment. It is performed by combining the exposure assessment with dose–response information. From a receptor's perspective, the risk from exposure to any chemical also depends on the potency associated with the effects and the duration of the exposure. The summary effects of the uncertainties in the preceding steps should also be included in this analysis.

The quantitative estimate of the risk is of principal interest to the regulatory agency or risk manager making a decision. The risk manager must consider the results of the risk characterization when evaluating the economics, societal aspects, and various benefits of the assessment. Factors such as societal pressure, technical uncertainties, and severity of the potential hazard influence how the decision makers respond to the risk assessment. As one might suppose, there is room for improvement in this step of the risk assessment.[5,6]

A risk estimate indicates the likelihood of occurrence of the different types of health or environmental effects in exposed populations. Risk assessment should include both human health and environmental evaluations (e.g., impacts on ecosystems). Ecological impacts include actual and potential effects on plants and animals (other than domesticated species). The number produced from the risk characterization, often representing the probability of adverse health effects being caused, must be carefully interpreted.

HAZARD RISK ASSESSMENT/ANALYSIS

There are several steps in evaluating the risk of an accident (see Figure 2.2). A more detailed figure is presented in Figure 2.3. If the system in question is a chemical plant, the specific steps to be followed in the risk evaluation process are listed here:

1. A brief description of the equipment and chemicals used in the plant is needed.
2. Any hazard in the system has to be identified. Hazards that may occur in a chemical plant include
   a. Fire
   b. Toxic vapor releases
   c. Slippage
   d. Corrosion

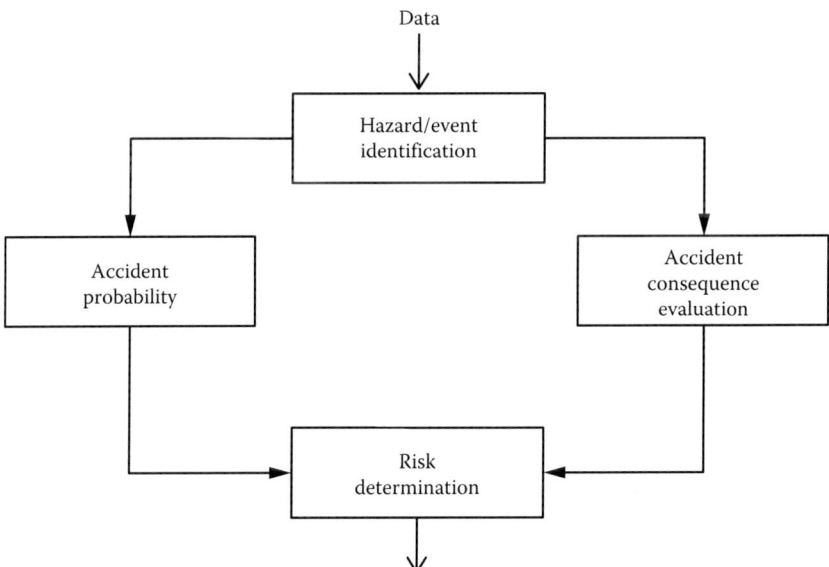

FIGURE 2.2   Simplified HZRA flowchart.

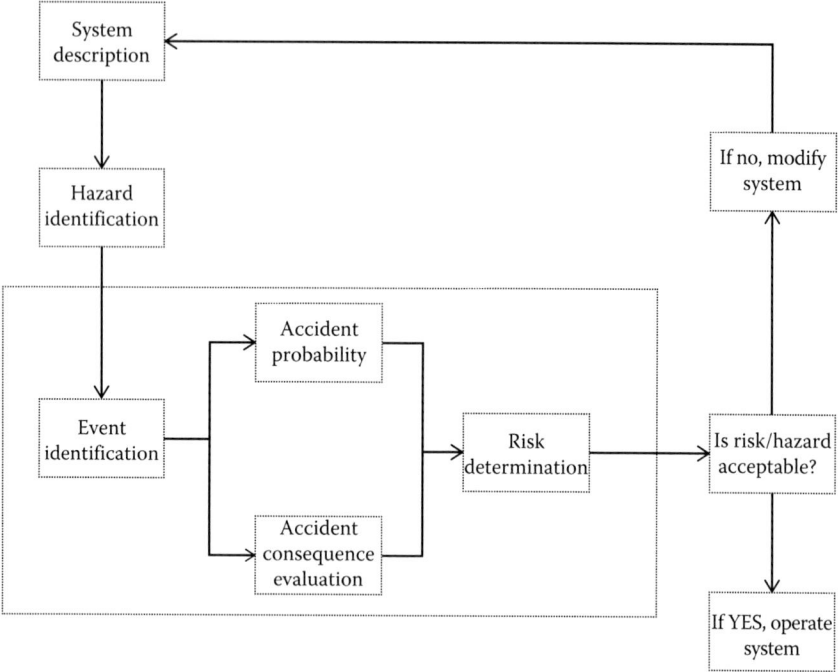

FIGURE 2.3   HZRA flowchart for a chemical plant.

    e. Explosions
    f. Rupture of pressurized vessels
    g. Runaway reactions
3. The event (or series of events) that will initiate an accident has (have) to be identified. An event could be a failure to follow correct safety procedures, improperly repaired equipment, or failure of a safety mechanism.
4. The probability that the accident will occur has to be determined. For example, if a chemical plant has a 10-year life span, what is the probability that the temperature in a reactor will exceed the specified temperature range over that lifetime? The probability can be ranked qualitatively from low to high. A low probability means that it is unlikely for the event to occur in the lifetime of the plant. A medium probability suggests that there is a possibility that the event will occur. A high probability means that the event will likely occur during the lifetime of the plant.
5. The severity of the consequences of the accident must be determined. This will be described in detail later in this section.
6. If the probability of the accident and the severity of its consequences are low, then the risk is usually deemed acceptable and the plant should be allowed to operate. If the probability of occurrence is too high or the damage to the surroundings is too great, then the risk is usually unacceptable and the system needs to be modified to minimize these effects.

As indicated in Figure 2.3, the heart of the HZRA approach is enclosed in the dashed box comprising Steps 3 through 6. The algorithm allows for re-evaluation of the process if the risk is deemed unacceptable (the process is repeated after system modification starting with Step 1).

Once again, it is important to note that an accident generally results from a sequence of events. Each individual event, therefore, represents an opportunity to reduce the frequency, consequence, and/or risk associated with the accident culminating from the individual events.

Hazard/Event Problem Identification

Hazard or event identification provides information on situations or chemicals and their releases, which can potentially harm the environment, life, or property. Information that is required to identify chemical hazards includes chemical identities; quantities and location of the chemicals in question; and, chemical properties such as boiling points, ignition temperatures, and toxicity to humans.[7] Obviously, the key word in this section is "identify," and the identification subject can be assisted by reviewing the following topics:

1. Process equipment
2. Classification of accidents
3. Fires, explosions, and hazardous spills

The next paragraph addresses hazard/event evaluation techniques and covers a number of methods used to identify some of the hazards common to the chemical process and manufacturing industries.

Generally, the hazard in question will take the form of either a chemical release or a "disaster arising from a blast/fire fragment problem," and some of the hazard/event evaluation techniques for these sorts of accidents include[3]

1. System checklists
2. Safety review/safety audit
3. "What if" analysis
4. Preliminary hazard analysis (PHA)
5. Hazard and operability (HAZOP) studies

Hazard/Event Probability

There are a host of reasons why accidents occur in industry. The primary causes are mechanical failure, operational error (human error), process upsets, and design errors. Keep in mind that the word *cause* has come to mean different things to different people. There are three steps that normally lead to an accident:

1. Initiation
2. Propagation
3. Termination

The chemical industry today is involved in a broad spectrum of manufacturing processes that range from biological preparations to the manufacturing of plastics and explosives. Although the basic plans and designs for these processes may be similar, each individual plant will have its own unique set of potential hazards.

The basic chemical processes that are common in industry include acylation, alkaline fusion, alkylation, amination, aromatization, calcination, carboxylation, causticization, combustion, condensation, coupling, cracking, diazotization, electrolysis, esterification, fermentation, halogenation, hydroforming, hydrolysis, isomerization, neutralization, nitration, nitrosation, oxidation, polymerization, pyrolysis, reduction, and thermal decomposition.[8] Some of these processes are considered to be more hazardous than others. Alkylation, amination, aromatization, combustion, diazotization, halogenation, nitration, oxidation, and polymerization are examples of the more hazardous processes. However, any process that exhibits one or more of the following characteristics should be considered to be extra hazardous:

1. The process is subject to an explosive reaction or detonation under normal conditions.
2. The process is subject to an explosive reaction or detonation when exposed to shock or abnormally high temperatures or pressure.
3. The process reacts violently with water.
4. The process is subject to spontaneous polymerization.
5. The process is subject to spontaneous heating.
6. The process is subject to exothermic reactions with the development of excessive temperatures and pressure.
7. The process normally operates at very high pressures or temperatures and may result in the massive release of flammable gases or vapors.
8. The process operates in or near the explosive range of the reactants or products.
9. The process is subject to a dust or mist explosion.
10. The process emits environmental contaminants or pollutants into the atmosphere.
11. The process uses or produces toxic substances.
12. The process uses or produces very corrosive materials.
13. The process emits dangerous radiation.
14. The process presents waste disposal problems.

Deviations from normal process conditions, as manifested by the following circumstances, must be well understood if accidents are to be prevented:

1. Abnormal (excursion) temperatures
2. Abnormal (excursion) pressures
3. Material flow stoppage
4. Equipment leaks or spills
5. Failure of equipment

Chemical processing under "extreme conditions" of high temperatures and pressure requires a more thorough analysis and extra safeguards. Explosions at higher initial temperatures and pressure are much more severe. Therefore, chemical processes under extreme conditions also require specialized equipment design and fabrication. Other factors that should be considered when evaluating a chemical process are the rate and order of the reaction, stability of the reaction, and the potential human health problems caused by the raw materials used.

Under the above circumstances, there is a need for high standards in equipment design, operation, and maintenance. Regardless of adequate safeguards and control of highly technological processes, accidents do and will continue to occur. It is therefore important to examine the causes of such accidents associated with specific pieces of equipment, supporting systems, and materials being handled. As noted earlier, the sequence of events resulting in an accident can generally be traced back to one or a combination of the following causes:

1. Equipment failure
2. Control system failure
3. Utilities and ancillary equipment outage
4. Human error
5. Fire exposure/explosions
6. Natural causes
7. Plant layout

HAZARD/EVENT CONSEQUENCES

Consequences of accidents can be classified qualitatively by the degree of severity. Factors that help to determine the degree of severity for chemicals are the concentration at which the hazard is released, the relative toxicity of the hazard, and, in the case of a chemical release, the length of time that a person is exposed to the hazardous agent. From a qualitative perspective, the worst-case consequence of a scenario is defined as a conservatively high estimate of the severity of the accident identified.[9] On this basis, one can qualitatively rank the consequences of accidents into *low*, *medium*, and *high* degrees of severity.[10] A low degree of severity means that the hazard consequence is nearly negligible, and the injury to persons, property, or the environment may be observed only after an extended period of time. The degree of severity is considered to be medium when the accident is serious but not catastrophic. An example of this could include a case where there is a release of a low concentration of a chemical that is considered to be highly toxic. Another example of a medium degree of severity could be a highly concentrated release of a less toxic chemical, large enough to cause injury or death to persons and damage to the environment unless immediate action is taken. There is a high degree of severity when the accident is catastrophic or the concentrations and toxicity of a chemical hazard are large enough to cause injury or death to many individuals, and there is long-term damage to the surrounding environment.

Potential consequences of other specific hazard/accident conditions can include

1. Flying shrapnel
2. Rocketing tank parts

3. Fireballs created by mechanically atomized drops of burning liquid and vapor
4. Secondary fires and explosions caused by flaming tank contents
5. Release of toxic or corrosive substances to the surroundings

Hazard Risk Characterization

Risk characterization often involves a judgment of the probability and severity of consequences based on the history of previous incidents, local experience, and the best available current technological information. It provides an estimation of

1. The likelihood (probability) of an accidental release based on the history of current conditions and controls at a facility, consideration of any unusual environmental conditions (e.g., areas on floodplains), or the possibility of simultaneous emergency incidents (e.g., flooding or fire hazards resulting in the release of hazardous materials)
2. The severity of consequences of human injury that may occur (acute, delayed, and/or chronic health effects), the number of possible injuries and deaths, and the associated high-risk groups
3. The severity of consequences on critical facilities (e.g., hospitals, fire stations, police departments, and communication centers)
4. The severity of consequences of damage to property (temporary, repairable, and permanent)
5. The severity of consequences of damage to the environment

The risk characterization process also attempts to attach meaning to a risk that has been calculated, including such factors as economic, social, and technological, plus selecting a course of action concerning the risk. Finally, interpreting and communicating the calculated risk value(s) to the public can be accomplished in a number or ways including the following:

1. Comparing with other known risks
2. Providing a perspective on the frequency or occurrence(s) of the risk
3. Explaining the sensitivity of the risk results and calculations to changes in input model data and scenario assumptions

The risk management process generally involves selecting a course of action that best addresses the risk in question and can include the following:

1. A cost–benefit analysis of risk
2. Measuring public perception
3. Determining acceptable levels of risk

Because so many hazards exist in everyday life, risk assessment must be used as a tool for evaluating those that are the most pressing or most hazardous. Over time, one may find that some activities are more hazardous than once perceived (i.e., smoking cigarettes or manufacturing polychlorinated biphenyls [PCBs]). Once the evidence is evaluated, these practices may be either stopped or limited. An assessment on an unknown chemical hazard or potentially unsafe practice attempts to predict what the consequences might be without waiting for final proof of an adverse impact.

Risk characterization estimates the risk associated with a process under investigation. The result of this characterization is a number that represents the probability of adverse effects from that process and/or from a substance released from that process. For instance, a risk characterization for all effects from a nuclear power plant might be expressed as one additional cancer case per one million people.

Once a risk characterization is made, the meaning of that risk must be evaluated. Public environmental and health agencies generally only consider risk greater than 10 in 1 million (1×10^{-5} or 10×10^{-6}) to be significant risks warranting action.

The major types of risks include (also see Chapter 1) the following:

1. *Individual risk*: This provides a measure of the risk to a person in the vicinity of a hazard/accident, including the nature of the injury or other undesired outcomes, and the likelihood of its occurrence. Individual risk is generally expressed in terms of a likelihood or probability or a specified undesired outcome per unit of time. For example, the individual risk of a fatality at a particular location near a nuclear power plant might be expressed as 1 in 100,000 per year or 10^{-5} per year. The risk to a person at a particular location depends on the probability of occurrence of the hazard event and on the probability of an adverse impact at that location should the event occur.

2. *Maximum individual risk (MIR)*: This is the maximum risk to an individual. The individual is considered to have a 70-year lifetime of exposure to a process or a chemical. For a discharge from a stack, for instance, the individual is considered to live downwind of the stack, never leaving this spot for even an hour, every day of a 70-year life span. The MIR is the risk to an individual subjected to this worst case.

3. *Population risk (PR)*: This is the risk to a population as a whole, expressed as a given number of deaths per thousand or per million people potentially exposed to the hazard.

4. *Societal risk*: This represents a measure of the risk to a group of people, including the risk of incidents potentially affecting more than one person. Individual risk is generally not significantly affected by the number of people involved in an incident.

5. *Risk indices*: A risk index is a single number measure of the risk associated with a facility. Some risk indices are qualitative or semi-quantitative, ranking risks in various general categories. Risk indices may also be a quantitative average or benchmarks based on other risk measures.

RISK UNCERTAINTIES/LIMITATIONS

The general subject of uncertainties/limitations is discussed in this section. The approach will examine the topic by briefly reviewing health risk and hazard risk concerns separately with uncertainties/limitations. However, before proceeding to these two topics, it is necessary to once again define and differentiate between accuracy and precision, and variability and uncertainty. The accuracy of a measurement or a calculation refers to that value relative to the correct value while precision relates to the significant figures of that value. Variability refers to the variables in the values that arise due to a calculation or measure while uncertainty, the concern at hand, refers to variations that occur due to the operations or process itself. For example, in measuring the diameter of a heat exchanger tube in a heat exchange, variations occur with repeated measurements of the tube's diameter while uncertainty arises from the measurements of many tubes (in the population).

HEALTH RISK

Although great controversy can surround results of risk assessments, especially quantitative risk assessments, they are useful in particular applications. They can help establish priorities for regulatory action or intervention of any type. A uniform risk assessment performed across a range of substances can create a spectrum of their health risks to humans. The limits of risk assessment can also be tested when government agencies (faced with the absence of other types of data and the need for action) must rely on risk assessment methods to establish health-based standards or guidelines to prevent human exposure to hazardous substances. Because of risk assessment shortcomings and the desire for greater specificity in measuring exposure, increasing interest is being shown in understanding pathologic changes at the molecular level with the hope that these investigations will lead to toxicological and epidemiological analyses of greater accuracy and sensitivity than that are currently available.[11,12] In a general sense, problems in this area arise because of

1. Uncertainty associated with available data
2. Concerns associated with assumed information
3. Uncertainty associated with governing equations
4. Concerns associated with limited and/or constrained governing equations
5. Concerns associated with overall analysis quality

In the risk characterization steps of a human health risk assessment, conclusions about health and dose–response are integrated with those from the exposure assessment step. In addition, confidence about these conclusions, including information about the uncertainties associated with each aspect of the assessment in the final risk summary, should be highlighted. In the previous assessment steps and in the risk characterization, the risk assessor should also distinguish between variability and uncertainty.

Variability arises from true heterogeneity in characteristics such as dose–response differences within a population or differences in contaminant levels in the environment. The values of some variables used in an assessment often change with time and space or across the population whose exposure is being estimated. Assessments should address the resulting variability in doses received by members of the target population. Individual exposure, dose, and risk can vary widely in a large population.

The central tendency and high-end individual risk descriptors are intended to capture the variability in exposure, lifestyles, and other factors that lead to a distribution of risk across a population.

Uncertainty, on the other hand, represents lack of knowledge about factors such as adverse effects or contaminant levels, which may be reduced with additional study. Generally, risk assessments involve several categories of uncertainty, and each merits consideration. Measurement uncertainty refers to the usual error that accompanies scientific measurements—standard statistical techniques applied to analytical quality control data can often be used to express measurement uncertainty. A substantial amount of uncertainty is often inherent in environmental sampling, and assessments should also address these uncertainties. Similarly, there are uncertainties associated with the use of scientific models, for example, dose–response models, and models of environmental fate and transport. Evaluation of model uncertainty should consider the scientific basis for the model and its available empirical validation.

It should be noted that there is no completely satisfactory way to generate accurate risk data since it is an inexact science fraught with uncertainties. At the very least, risk characterization should be checked against experience for reasonableness since the size and quality of the data employed does not permit an accurate quantitative estimate with a high degree of confidence. Careful documentation of all four parts of a risk assessment should also be maintained to prevent the practitioner from falling into traps that can influence the final results or pass the risk via a cross-media process onto another location or vulnerable population. The author believes that the EPA and OSHA have compounded the human health risk assessment uncertainty problem by some of their ambiguous and conflicting rules and regulations.

Finally, it should also be noted that less information is available on the similarities or differences in the degree of response of experimental animals as compared to humans to varying doses of a chemical. In these tests, the animals are, out of necessity, administered high doses of the chemical whereas humans are usually exposed to much lower levels. This makes it necessary to extrapolate from results seen at high doses to the results expected at low doses in humans. The validity of these extrapolations is, in most cases, not amenable to experimental verification. Thus, while the test species may serve as an approximate measure of the potential of a chemical to cause toxic effects in humans, attempts to quantify human risk on the basis of such studies remain subject to *considerable* scientific uncertainty. This uncertainty is particularly critical when attempts are made to predict carcinogenic responses in humans using data from tests in rats and mice.

HAZARD RISK

The reader should note that this topic was discussed earlier. Estimating the magnitude of risks that cannot be measured accurately or directly often requires employing assumptions that cannot be verified or tested experimentally. Obviously, knowledge about the present and the future is never completely accurate. Inadequate knowledge is usually the largest cause of uncertainty. The inadequacy of knowledge means that the full extent of the uncertainty is also unknown. Uncertainty due to variability occurs when a single number (as often employed in risk analysis) is used to describe something that truly has multiple or variable values. Variability is often ignored by using values based on the mean of all the values occurring within a group. Information on sources of uncertainties and limitations of input data are available.[13] Some of this material is provided here.

Uncertainties and limitations in system description data could include the following:

1. Process description or drawings are incorrect or out of date.[4]
2. Procedures do not represent actual operations.
3. Site area maps and population data may be incorrect or out of date.
4. Weather data from the nearest available site may be inappropriate due to its distance from the site or dissimilarity to microclimatic conditions.

Hazard identification data could have uncertainties and/or limitations because:

1. Recognition of major hazards may be incomplete.
2. Screening techniques employed for the selection of hazards for further evaluation may omit important cases.

Frequency techniques may have sources of uncertainty or limitation due to

1. *Uncertainties in modeling* due to
   a. Extrapolation of historical data to larger-scale operations that may overlook hazards introduced by scaling-up to larger equipment
   b. Limitation of fault tree theory that requires system simplification
   c. Incompleteness in fault and event tree analysis[14]
2. *Uncertainties in data* that may be caused because
   a. Data may be inaccurate, incomplete, or inappropriate
   b. Data from related activities might not be directly applicable
   c. Data generated by expert judgment may be inaccurate
   d. Characterization of the general population is improper or incomplete

Consequence techniques may have sources of uncertainty or limitations due to calculational burdens (even with computers). For example, they may arise from a number of dispersion modeling variables for chemical accidents, including

1. *Uncertainties in physical modeling* due to
   a. Inappropriate model selection
   b. Incorrect or inadequate physical basis for model
   c. Inadequate validation
   d. Inaccurate model parameters
2. *Uncertainties in physical model data* due to
   a. Input data (composition, temperature, pressure)
   b. Source terms for dispersion and other models

3. *Uncertainties in effects modeling* due to
   a. Animal data that may be inappropriate for humans (especially for toxicity)
   b. Omission of mitigating effects
   c. Lack of epidemiological data on humans of the same sex, age, education, etc.

Risk estimation may have sources of uncertainty or limitation due to

1. *Assumptions of symmetry* such as
   a. Uniform wind roses that rarely occur
   b. Uniform ignition sources that may be incorrect
   c. Single point source for all incidents that may be inaccurate
2. *Assumptions to reduce the complexity of the analysis* such as
   a. A single condition of wind speed and stability that may be too restrictive
   b. A limited number of ignition cases that can reduce accuracy
   c. General problems with the quality of data

The reader should note that since many risk analyses have been conducted on the basis of fatal effects, there are also uncertainties on precisely what constitutes a fatal dose of thermal radiation, blast effect, a toxic chemical, and so on.

ILLUSTRATIVE EXAMPLES

Illustrative Example 2.1

What are the two concepts that generally arise in a discussion associated with uncertainty?

Solution

Generally, uncertainty consists of two parts: variability and inadequate knowledge. Uncertainty due to variability occurs when a single number is used to describe something that truly has multiple or variable values. Variability is often ignored by using values based on the mean of all the values occurring within a group. A second type of variability is when a single value exists but constantly changes over time.

Despite the importance of variability of data, inadequate knowledge is usually the largest cause of uncertainty. Three common sources of inadequate knowledge include:

1. Parameter uncertainty or lack of knowledge of accurate parameter values due to measurement errors, random errors, systematic errors, etc.
2. Model uncertainty due to errors arising from incorrect concepts of reality and the use of incorrect models for describing chemical releases, transport, and exposure
3. Decision rule uncertainty or lack of knowledge regarding how best to interpret modeling outcomes and resulting consequences

Illustrative Example 2.2

Discuss health problems with delayed effects and their impact on risk estimates.

Solution

The cause of a health problem may not be obvious if the health problem effect is delayed. This applies to many chronic diseases. The question of causality is important for this class of risk. The delayed effect obviously increases the uncertainties associated with any risk estimate.

Illustrative Example 2.3

Can a health risk characterization provide information on exactly what to do about a specific problem?

Solution

No. Risk characterization is often imprecise in that it draws upon available information about a problem, applies scientific principles, and then provides guidance on potential risk. It does help identify hazards. How that information is used to decide what steps, if any, to take to reduce the hazard is not part of the risk characterization process.

Illustrative Example 2.4

List some of the types of information a hazard risk assessment can provide.

Solution

1. Identification and description of hazards and accident events that could lead to undesirable consequences
2. A qualitative estimate of the likelihood and consequence of each accident event sequence
3. A relative ranking of the risk of each hazard and accident event sequence
4. Some suggested approaches to risk reduction

For the chemical process industry, these results are normally provided to plant management and engineering or research groups, as appropriate, so that overall plant and process safety can be improved, and both on- and off-site risks can be minimized.

Illustrative Example 2.5

The word "what" appears in numerous hazard identification procedures. List some questions/comments that are related to this term.

Solution

The word "what" is used to ascertain conditions or connections that may exist for a specific piece of equipment or process in a manufacturing facility. Some of the "is" and "is nots" related to conditions or connections of a system that can be related to this question of "what" are summarized as follows:

	Is	**Is not**
What	What do we know?	What do we not know?
	What was observed?	What was not observed?
	What is a related problem?	What is unrelated?
	What are the constraints?	What is not a constraint?
	What is expected?	What is unexpected?
	What is the same?	What is different?
	What is important?	What is not important?
	What resources are needed?	
	What are the criteria?	
	What is the purpose?	

Illustrative Example 2.6

List and briefly discuss human errors that can occur in a chemical processing plant that can lead to or cause accidents.

Solution

Some examples of human error that can occur in a chemical processing plant are

1. *Design errors*: Improper design of plant-specific processes and improper sizing and specifications for plant equipment, controls, etc.
2. *Construction errors*: Poor construction techniques or low-quality construction materials
3. *Procedural errors*: Not following proper operating and/or maintenance procedures due to poor or inadequate training
4. *Management errors*: Lack of attention to worker training or performance and lack of management expectations for excellence in worker and plant performance
5. *Maintenance errors*: Lack of detail in maintenance requirements, scheduling, and safety procedures due to both worker and management errors

Illustrative Example 2.7

Discuss how hurricanes can lead to industrial accidents.

Solution

High winds, torrential rain, extreme high tides, lightning, and occasional tornadoes, which are all associated with hurricanes, can cause accidents to occur in chemical plants, refineries, utilities, offshore oil wells, and so on. An HZRA can determine if adequate control, containment, safety, and so on, can be maintained during such an event.

Illustrative Example 2.8

Describe the differences between catastrophic failure, degraded failure, and incipient failure.

Solution

The differences in failures can be defined as follows:

Catastrophic failure—A sudden failure where one or more fundamental functions have been terminated.

Degraded failure—A failure that occurs over time. Unlike catastrophic failure, a degraded failure does not terminate a function; it only impedes a function. This may result in undesired output and overtime at a plant if not fixed. It will eventually become a catastrophic failure if not corrected.

Incipient failure—A failure that can result, if left unattended, in a catastrophic failure due to improper design of the equipment.

Illustrative Example 2.9

Discuss the problems in valuing life relative to characterizing risk.

Solution

There are various estimates of the value of life, ranging at the high end from about $2 million per life to about $200,000 per life. The choice of value in this range depends a great deal on the ethical basis upon which the estimates are made. Beyond being seemingly callous in putting a "price tag" on a person's life, the decision to value one life more than another subjects the risk assessment to potential criticism of racism, bias, insensitivity, subjectivity, and so on. When attempting to develop a cost–benefit analysis involving risk of human fatalities, it is essential that life valuation be done as objectively as possible, and with as much public input and involvement as is reasonably possible to ensure that the process is unbiased, fair, and equitable.

REFERENCES

1. L. Theodore, J. Reynolds, and K. Morris, *Health, Safety and Accident Prevention: Industrial Applications*, Theodore Tutorials (originally published by USEPA, RTP, NC), East Williston, NY, 1996.
2. M.K. Theodore and L. Theodore, *Introduction to Environmental Management*, CRC Press/Taylor & Francis Group, Boca Raton, FL, 2010.
3. L. Theodore and R. Dupont, *Environmental Health Risk and Hazard Risk Assessment: Principles and Calculations*, CRC Press/Taylor & Francis Group, Boca Raton, FL, 2013.
4. G. Burke, B. Singh, and L. Theodore, *Handbook of Environmental Management and Technology*, 2nd edn., John Wiley & Sons, New York, NY, 2000.
5. D. Paustenback, *The Risk Assessment of Environmental and Human Health Hazards: A Textbook of Case Studies*, John Wiley & Sons, Hoboken, NJ, 1989.
6. G. Masters, *Introduction to Environmental Engineering and Science*, Prentice Hall, Upper Saddle River, NJ, 1991.

7. J. Santoleri, J. Reynolds, and L. Theodore, *Introduction to Hazardous Waste Incineration*, 2nd edn., John Wiley & Sons, Hoboken, NJ, 2000.

8. Engineering and Safety Service, *Hazard Survey of the Chemical and Allied Industries*, Technical Survey No. 3, American Insurance Association, New York, NY, 1979.

9. AIChE, *Guidelines for Hazard Evaluation Procedures*, Center for Chemical Process Safety, New York, NY, 1992.

10. Adapted from U.S. EPA, *Technical Guidance for Hazard Analysis, Emergency Planning for Extremely Hazardous Substances*, EPA-OSWER-88-0001, Office of Solid Waste and Emergency Response, Washington, DC, 1987.

11. P. Shields and N. Hanes, Molecular epidemiology and the genetics of environmental cancer, *JAMA*, 266, 681–687, 1991.

12. U.S. EPA, *Unfinished Business: A Comparative Assessment of Environmental Problems, Overview Report*, EPA-230-2-87-025a, Office of Policy, Planning and Evaluation, Washington, DC, 1987.

13. L. Theodore, personal notes, East Williston, NY, 2001.

14. S. Shaefer and L. Theodore, *Probability and Statistics Applications for Environmental Science* (adapted from), CRC Press, Boca Raton, FL, 2007.

3 Introduction to Probability and Statistics

INTRODUCTION

As noted in the introductory chapter, this chapter addresses the key fundamentals and principles associated with both *probability* and *statistics*. Webster[1] defines probability as "the number of times something will probably occur over the range of possible scenarios, expressed as a ratio." Most applied scientists, including engineers, would claim that probability is concerned with describing the phenomena of chance and randomness. Statistics is an order of magnitude more difficult to define. Webster[1] provides the following: "facts or data of a numerical kind, assembled, classified, and tabulated so as to present significant information about a given subject; the science of assembling, classifying, tabulating, and analyzing such facts or data." There are obviously many other definitions.

Generally, *statistics* can be simply defined as the branch of science which deals with collecting, arranging, and using numerical facts or data arising from natural phenomena or experiments. Thus, *statistic* is an item of information deduced from the application of statistical methods.

The most frequently encountered statistical problems are those which involve one or more of the following features:

1. Reduction of data
2. Relationships between two or more variables
3. Estimates and tests of significance
4. Reliability of inferences depending on one or more variables
5. Analysis of the fluctuations in a measurable quantity which can arise naturally during the course of a particular operation[2]

No attempt is made in this chapter to introduce and apply packaged computer programs that are presently available; the emphasis is to provide the reader with an understanding of the fundamental principles so that he or she learns how statistical methods can be used in environmental risk analysis. The bulk of the material presented in this chapter is drawn from Theodore and Taylor.[3]

The sections in this chapter address the following five subject areas:

1. Probability Definitions and Interpretations
2. Basic Probability Theorems
3. Median, Mean, and Standard Deviation
4. Random Variables
5. Illustrative Examples

PROBABILITY DEFINITIONS AND INTERPRETATIONS

Probabilities are nonnegative numbers associated with the outcomes of the so-called random experiments. A random experiment is an experiment whose outcome is uncertain. Examples include throwing a pair of dice, tossing a coin, counting the number of defectives in a sample from a lot of manufactured items, and observing the time to failure of a tube in a heat exchanger, a seal in a pump, or a bus section in an electrostatic precipitator. The set of possible outcomes of a random experiment is called the *sample space* and is usually designated by S. The probability of event A, $P(A)$, is the sum of the probabilities assigned to the outcomes constituting the subset A of the sample space S.

Consider, for example, tossing a coin twice. The sample space of heads (H) and tails (T) can be described as

$$S = \{HH, HT, TH, TT\} \tag{3.1}$$

If probability ¼ is assigned to each element of S, and A is the event of at least one head, then

$$A = (HH, HT, TH) \tag{3.2}$$

The sum of the probabilities assigned to the elements of A is ¾. Therefore, $P(A) = ¾$.

The description of the sample space is not unique. The sample space S in the case of tossing a coin twice could be described in terms of the number of heads obtained. Then,

$$S = (0, 1, 2) \tag{3.3}$$

Suppose probabilities ¼, ½, ¼ are assigned to the outcomes 0, 1, and 2, respectively. Then, A, the event of at least one head, would have for its probability,

$$P(A) = P\{1, 2\} = ¾ \tag{3.4}$$

How probabilities are assigned to the elements of the sample space depends on the desired interpretation of the probability of an event. Thus, $P(A)$ can be interpreted as *theoretical relative frequency*—that is, a number about which the relative frequency of event A tends to cluster as n, the number of times the random experiment is performed, increases indefinitely. This is the objective interpretation of probability. Under this interpretation, saying $P(A)$ is ¾ in the example given above means that if a coin is tossed twice n times, the proportion of times one or more heads occur clusters about ¾ as n increases indefinitely.

As another example, consider a single valve that can stick in an open (O) or closed (C) position. The sample space can be described as follows:

$$S = \{O, C\} \tag{3.5}$$

Suppose that the valve sticks twice as often in the open position as it does in the closed position. Under the theoretical relative frequency interpretation, the probability assigned to element O in S would be 2/3, twice the probability assigned to element C. If two such valves are observed, the sample space S can be described as

$$S = \{OO, OC, CO, CC\} \tag{3.6}$$

Assuming that the two valves operate independently, a reasonable assignment of probabilities to the elements of S as listed could be 4/9, 2/9, 2/9, and 1/9. The reason for this assignment will become clear after consideration of the concept of independence. If A is the event of at least one valve sticking in the closed position, then

$$S = \{OC, CO, CC\} \tag{3.7}$$

The sum of the probabilities assigned to the elements of A is 5/9. Therefore, $P(A) = 5/9$.

Probability P/A can also be interpreted subjectively as a measure of degree of belief, on a scale from 0 to 1, that the event A occurs. This interpretation is frequently used in ordinary conversation. For example, if someone says, "The probability that I will go to the racetrack today is 90%," then 90% is a measure of the person's belief that he will go to the racetrack. This interpretation is also used when, in the absence of concrete data needed to estimate an unknown probability on the basis of observed relative frequency, the personal opinion of an expert is sought. For example, an expert might be asked to estimate the probability that the seals in a newly designed pump will leak at high pressures. The estimate would be based on the expert's familiarity with the history of pumps of similar design.

Various combinations of the occurrence of any two events A and B can be indicated in set notations as follows:

$\overline{A}$ A does not occur

$\overline{B}$ B does not occur

$A + B$ A occurs or B occurs in the mutually inclusive sense to indicate the occurrence of A, B, or both A and B

AB A occurs and B occurs

$A\overline{B}$ A occurs and B does not occur

Venn diagrams (Figure 3.1) provide a pictorial representation of these events. In set terminology, $\overline{A}$ is called the *complement* of A, that is, the set of elements in S that are not in A. An alternate notation for the *complement* of A is A^c or A'. AB is called the *intersection* of A and B, that is, the set of elements in both A and B. An alternate notation in the literature for the intersection of A and B is $A \cap B$. $A + B$ is called the *union* of A and B, that is, the set of elements A, B, or both A and B. An alternate notation for the union of A and B is $A \cup B$.

When events A and B have no elements in common they are said to be *mutually exclusive*. A set having no elements is called the *null set* and is usually designated by ϕ.

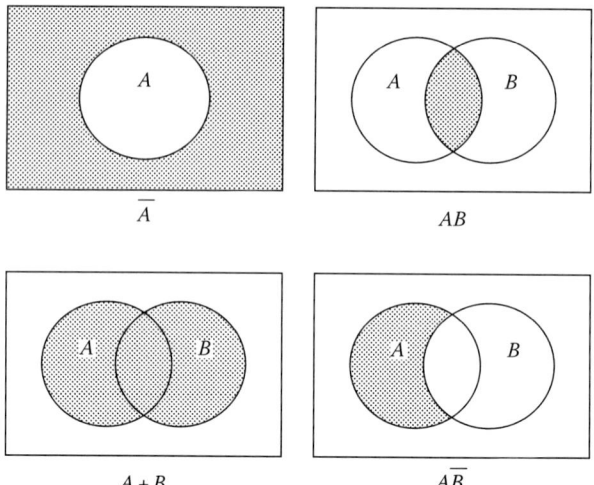

FIGURE 3.1 Venn diagrams.

Thus, if events A and B are mutually exclusive, then $AB = \phi$. Note that the union of A and B consists of three mutually exclusive events: $A\bar{B}, AB, \bar{A}B$.

The algebra of sets—Boolean algebra—governs the way in which sets can be manipulated to form equivalent sets. The principal Boolean algebra laws used for this purpose are as follows.

Commutative law:

$$A + B = B + A \tag{3.8}$$

$$AB = BA \tag{3.9}$$

Associative law:

$$(A + B) + C = A + (B + C) = A + B + C \tag{3.10}$$

$$(AB)C = A(BC) = ABC \tag{3.11}$$

Distributive law:

$$A(B + C) = AB + AC \tag{3.12}$$

$$A + BC = (A + B)(A + C) \tag{3.13}$$

Absorption law:

$$A + A = A \tag{3.14}$$

$$AA = A \tag{3.15}$$

De Morgan's law:

$$\overline{(A + B)} = \bar{A}\bar{B} \tag{3.16}$$

$$\overline{AB} = \bar{A} + \bar{B} \tag{3.17}$$

If the letter symbols for sets are replaced by numbers, the commutative and associative laws become familiar laws of arithmetic. In Boolean algebra, the first of the two distributive laws, Equation 3.12, has an analogous counterpart in arithmetic. The second, Equation 3.13, does not.

In risk analysis, Boolean algebra is used to simplify expressions for complicated events. For example, consider the event

$$T = (A + B + C)(\bar{C} + D)(E + F) \tag{3.18}$$

where

$$\bar{C} = D + E$$

$$\bar{D} = A + B$$

Note that $C\bar{C} = \phi$, the null set.

After substituting

$$\bar{C} = D + E \tag{3.19}$$

and noting that

$$AD + AD = AD$$
$$BD + BD = BD \tag{3.20}$$
$$AD + BD = \phi$$

the expression for T in Equation 3.18 reduces to

$$T = \left(AE + BE + CD\right)\left(E + F\right) \tag{3.21}$$

Noting that

$$T = (AE + BE + CD)(E + F)$$

gives:

$$T = AE + AEF + BE + BEF \tag{3.22}$$

This result indicates that the occurrence of any one of the events in the right member of Equation 3.22 results in the occurrence of event T.

BASIC PROBABILITY THEOREMS[3]

The mathematical properties of $P(A)$, the probability of event A, are deduced from the following postulates governing the assignment of probabilities to the elements of a sample space, S.

1. $P(S) = 1$
2. $P(A) \geq 0$ for any event A
3. If $A_1, \ldots, A_n, \ldots$ are mutually exclusive, then
   $$P(A_1 + A_2 + \cdots + A_n + \cdots) = P(A_1) + P(A_2) + \cdots + P(A_n) + \cdots$$

In the case of a discrete sample space (i.e., a sample space consisting of a finite number or countable infinitude of elements), these postulates require that the numbers assigned as probabilities to the elements of S be nonnegative and have a sum equal to 1. These requirements do not result in the complete specification of the numbers assigned as probabilities. The approved interpretation of probability must also be considered. The mathematical properties of the probability of any event are the same regardless of how this probability is interpreted. These properties are formulated in theorems logically deduced from the postulates given earlier without the need for appeal to interpretation. Three basic theorems are

Theorem 1: $\qquad\qquad P(\bar{A}) = 1 - P(A) \qquad\qquad$ (3.23)

Theorem 2: $\qquad\qquad 0 \leq P(A) \leq 1 \qquad\qquad$ (3.24)

Theorem 3: $\qquad\qquad P(A + B) = P(A) + P(B) - P(AB) \qquad\qquad$ (3.25)

Theorem 1 states that the probability that A does not occur is one minus the probability that A occurs. Theorem 2 states that the probability of any event lies between 0 and 1. Theorem 3, the addition theorem, provides an alternative way of calculating the probability of the union of two events as the sum of their probabilities minus the probability of their intersection. The addition theorem can be extended to three or more events. In the case of three events A, B, and C, the addition theorem becomes

$$P(A+B+C) = P(A) + P(B) + P(C) - P(AB)$$
$$- P(AC) - P(BC) + P(ABC) \qquad (3.26)$$

For four events A, B, C, and D, the addition theorem becomes

$$P(A+B+C+D) = P(A) + P(B) + P(C) + P(D) - P(AB) - P(AC) - P(AC)$$
$$- P(AD) - P(BC) - P(BD) - P(CD) + P(ABC) + P(ABD)$$
$$+ P(BCD) + P(ACD) - P(ABCD) \qquad (3.27)$$

To illustrate the application of the three basic theorems (Equations 3.23 through 3.25), consider what happens when one draws a card at random from a deck of 52 cards. The sample space S may be described in terms of 52 elements, each corresponding to one of the cards in the deck. Assuming that each of the 52 possible outcomes would occur with equal relative frequency in the long run leads to the assignment of equal probability, 1/52, to each of the elements of S. Let A be the event of drawing an ace and B the event of drawing a club. Thus, A is a subset consisting of four elements, each of which has been assigned probability: 4/52, and $P(A)$ is the sum of these probabilities: 4/52. Similarly, the following probabilities are obtained:

$$P(B) = \frac{13}{52} \qquad P(AB) = \frac{1}{52}$$

Application of Theorem 1 gives

$$P(\bar{A}) = \frac{48}{52} \qquad P(\bar{B}) = \frac{39}{52}$$

Application of the addition theorem gives

$$P(A+B) = \frac{4}{52} + \frac{13}{52} - \frac{1}{52} = \frac{16}{52}$$

$P(A+B)$, the probability of drawing an ace or a club, can be calculated without using the addition theorem by calling $A+B$, the union of A and B, a set consisting of 16 elements. (The number, 16, is obtained by adding the number of aces, 4, to the number of clubs, 13, and subtracting the card that is counted twice—once as an ace and once as a club.) Since each of the 16 elements in $A+B$ has been assigned probability 1/52, $P(A+B)$ is the sum of the probabilities assigned, namely, 16/52.

The *conditional probability* of event B given A is denoted by $P(B|A)$ and defined as

$$P(B\,|\,A) = \frac{P(AB)}{P(A)} \tag{3.28}$$

where $P(B|A)$ can be interpreted as the proportion of A occurrences that also feature the occurrence of B.

For example, consider the random experiment of drawing two cards in succession from a deck of 52 cards. Suppose the cards are drawn *without* replacement (i.e., the first card drawn is not replaced before the second is drawn). Let A denote the event that the first card is an ace and B, the event that the second card is an ace. The sample space S can be described as a set of 52 times 51 pairs of cards. Assuming that each of these (52)(51) pairs has the same theoretical relative frequency, assign probability 1/(52)(51) to each pair. The number of pairs featuring an ace as the first and second card is (4)(3). Therefore,

$$P(AB) = \frac{(4)(3)}{(52)(51)}$$

The number of pairs featuring an ace as the first card and one of the other 51 cards as the second is (4)(51). Therefore,

$$P(A) = \frac{(4)(51)}{(52)(51)}$$

Applying the definition of conditional probability, Equation 3.28, yields

$$P(B\,|\,A) = \frac{3}{51} = 0.0588$$

as the conditional probability that the second card is an ace, given that the first is an ace. The same result could have been obtained by computing $P(B)$ on a new sample space consisting of 51 cards, three of which are aces. This illustrates the two methods for calculating a conditional probability. The first method calculates the conditional probability in terms of probabilities computed on the original sample space by means of the definition in Equation 3.28. The second method uses the given event to construct a new sample space on which the conditional probability is computed.

Conditional probability can also be used to formulate a definition for the independence of two events A and B. Event B is defined to be independent of event A if and only if

$$P(B\,|\,A) = P(B) \tag{3.29}$$

Similarly, event A is defined to be independent of event B if and only if

$$P(A\,|\,B) = P(A) \tag{3.30}$$

From the definition of conditional probability in Equation 3.28, one can deduce the logically equivalent definition that event A and event B are independent if and only if

$$P(AB) = P(A)P(B) \qquad (3.31)$$

To illustrate the concept of independence, consider again the random experiment of drawing two cards in succession from a deck of 52 cards. This time suppose that the cards are drawn *with* replacement (i.e., the first card is replaced in the deck before the second card is drawn). As before, let A denote the event that the first card is an ace, and B the event that the second card is an ace. Then,

$$P(B \mid A) = \frac{4}{52}$$

and since $P(B|A) = P(B)$, B and A are independent events.

From the definition of $P(B|A)$ and $P(A|B)$, one can deduce from Equation 3.28 the multiplication theorem,

$$P(AB) = P(A)P(B \mid A) \qquad (3.32)$$

$$P(AB) = P(B)P(A \mid B) \qquad (3.33)$$

The multiplication theorem provides an alternate method for calculating the probability of the intersection of two events.

The multiplication theorem can be extended to the case of three or more events. For three events A, B, C, the multiplication theorem states

$$P(ABC) = P(A)P(B \mid A)P(C \mid AB) \qquad (3.34)$$

For four events A, B, C, and D, the multiplication theorem states

$$P(ABCD) = P(A)P(B \mid A)P(C \mid AB)P(D \mid ABC) \qquad (3.35)$$

Consider now n mutually exclusive events $A_1, A_2, \ldots, A_n$, whose union is the sample space S. Let B be any given event. Then, Bayes' theorem states

$$P(A_i \mid B) = \frac{P(A_i)P(B \mid A_i)}{\sum_{i=1}^{n} P(A_i)P(B \mid A_i)}; \quad i = 1, \ldots, n \qquad (3.36)$$

where

$P(A_1)$, $P(A_2)$, $\ldots$, $P(A_n)$ are called the *prior probabilities* of $A_1, A_2, \ldots, A_n$

$P(A_1|B)$, $P(A_2|B)$ are called the *posterior probabilities* of $A_1, A_2, \ldots, A_n$

Bayes' theorem provides the mechanism for revising prior probabilities, that is, for converting them into posterior probabilities on the basis of the observed occurrence of some given event.

MEDIAN, MEAN, AND STANDARD DEVIATION[3]

One basic way of summarizing data is by the computation of a central value. The most commonly used central value statistic is the arithmetic average, or the mean. This statistic is particularly useful when applied to a set of data having a fairly symmetrical distribution. The mean is an efficient statistic in that it summarizes all the data in the set and because each piece of data is taken into account in its computation. The formula for computing the mean is

$$\bar{X} = \frac{X_1 + X_2 + X_3 + \cdots + X_n}{n} = \frac{\sum_{i=1}^{n} X_i}{n} \tag{3.37}$$

where
$\bar{X}$ is the arithmetic mean
X_i is any individual measurement
n is the total number of observations
$X_1, X_2, X_3, \ldots$ are the measurements 1, 2, and 3, respectively

The *arithmetic mean* is not a perfect measure of the true central value of a given data set. It can overemphasize the importance of one or two extreme data points. Many measurements of a normally distributed data set will have an arithmetic mean that closely approximates the true central value.

When a distribution of data is asymmetrical, it is sometimes desirable to compute a different measure of central value. The second measure, known as the *median*, is simply the middle value of a distribution, or the quantity above which half the data lie and below which the other half lie. If n data points are listed in their order of magnitude, the median is the $[(n+1)/2]$th value. If the number of data is even, then the numerical value of the median is the value midway between the two data nearest the middle. The median, being a positional value, is less influenced by extreme values in a distribution than the mean. However, the median alone is usually not a good measure of central tendency. To obtain the median, the data provided must first be arranged in order of magnitude.

Another measure of central tendency used in specialized applications is the *geometric mean*, $\bar{X}_G$. The geometric mean can be calculated using the following equation:

$$\bar{X}_G = \sqrt[n]{(X_1)(X_2)\cdots(X_n)} \tag{3.38}$$

The most commonly used measure of dispersion, or variability, of sets of data is the *standard deviation*, σ. Its defining formula is given by the expression:

$$\sigma = \sqrt{\frac{\sum \left(X_i - \overline{X} \right)^2}{n-1}} \tag{3.39}$$

where
σ is the standard deviation (always positive)
X_i is the value of the ith data point
$\overline{X}$ is the mean of the data sample
n is the total number of observations

The expression $\left(X_i - \overline{X} \right)$ indicates that the deviation of each piece of data from the mean is taken into account by the standard deviation. Although the defining formula for the standard deviation gives insight into its meaning, the following algebraically equivalent formula makes computation much easier (if now applied to the temperature, T):

$$\sigma = \sqrt{\frac{\sum \left(T_i - \overline{T} \right)^2}{n-1}} = \sqrt{\frac{n \sum \left(T_i^2 - \left(\sum T_i \right)^2 \right)}{n(n-1)}} \tag{3.40}$$

Variance is denoted as σ^2.

As noted before, statistics is often concerned with the extraction of information from observed data. A distinction between a sample and the population (the totality of measurements from which the sample was drawn) needs to be made in order to draw an inference about a population from the information contained in a sample. To assist in this distinction, a characteristic of a population is normally symbolized with a Greek letter, for example, $\mu = 36.7$ g, $\sigma^2 = 100$ g, and $\sigma = 10$ g. A characteristic of a sample is normally symbolized with an alphabet letter, for example, $\overline{x} = 24.3$ g, $s^2 = 81$ g, and $s = 9$ g.

Chebyshev's theorem provides a rather unique and interesting interpretation of the sample standard deviation as a measure of the spread (dispersion) of sample observations about their mean.[3] The theorem states that at least $(1 - 1/k^2)$, $k > 1$, of the sample observations lie in the interval $(\overline{X} - ks, \overline{X} + ks)$. For $k = 2$, for example, this means that at least 75% of the sample observations lie in the interval $(\overline{X} - 2s, \overline{X} + 2s)$. The smaller the value of s, the greater the concentration of observations in the vicinity of X.

In the case of a random sample of observations on a continuous random variable assumed to have a so-called normal pdf (see also Chapter 11), the graph of which is a bell-shaped curve, the following statements give a more precise interpretation of the sample standard deviation s as a measure of spread or dispersion.

1. $\overline{X} \pm s$ includes approximately 68% of the sample observations.
2. $\overline{X} \pm 2s$ includes approximately 95% of the sample observations.
3. $\overline{X} \pm 3s$ includes approximately 99.7% of the sample observations.

The source of these percentages is the normal probability distribution, which is studied later in more detail in Section III.

RANDOM VARIABLES

A random variable is a real-valued function defined over the sample space S of a random experiment. The domain of the function is S, and the real numbers associated with the various possible outcomes of the random experiment constitute the range of the function. If the range of the random variable consists of a finite number or countable infinitude of values, the random variable is classified as *discrete*. If the range consists of a non-countable infinitude of values, the random variable is classified as *continuous*. A set has a countable infinitude of values if they can be put into one-to-one correspondence with positive integers. The positive even integers, for example, consist of a countable infinitude of numbers. The even integer $2n$ corresponds to the positive integer n for $n = 1, 2, 3, \ldots$ The real numbers in the interval $(0, 1)$ constitute a non-countable infinitude of values.

Defining a random variable on a sample space S amounts to coding the outcomes in real numbers. Consider, for example, the random experiment involving the selection of an item at random from a manufactured lot. Associate $X = 0$ with the drawing of a nondefective item and $X = 1$ with the drawing of a defective item. Then, X is a random variable with range $(0, 1)$ and is therefore discrete.

Let X denote the number of the throw on which the first failure of a switch occurs. Then, X is also a discrete random variable with range $\{1, 2, 3, \ldots, n, \ldots\}$. Note that the range of X consists of a countable infinitude of values and that X is therefore discrete. Alternatively, suppose that X denotes the time to failure of a bus section in an electrostatic precipitator. Then, X is a continuous random variable whose range consists of the real numbers greater than zero.

The probability distribution of a random variable concerns the distribution of probability over the range of the random variable. The distribution of probability is specified by the pdf (*probability distribution function*). Chapter 4, is concerned with the general properties of the pdf for the case of discrete and continuous random variables. Special pdfs find extensive application in environmental risk analysis.

ILLUSTRATIVE EXAMPLES

Illustrative Example 3.1

Suppose that the failure of either a generator or a switch in a utility plant will cause interruption of electrical power. If the probability of a generator failure is 0.02, the probability of switch failure is 0.01, and the probability of both failing is 0.0002, what is the probability of power interruption?

Solution

If A denotes the event of generator failure, and B the event of switch failure, then $A + B$ denotes the event of power interruption, the probability of which is given by Equation 3.25.

$$P(A+B) = P(A) + P(B) - P(AB)$$
$$P(A+B) = 0.02 + 0.01 - 0.0002 = 0.0298$$

Illustrative Example 3.2

Consider the case of a box of 100 transistors from which a sample of 2 items is to be drawn *without* replacement. If the box contains 5 defective transistors, what is the probability that the sample contains exactly 2 defectives?

Solution

Let A denote the event that the first transistor drawn is defective, and B, the event that the second is defective. Then, the probability that the sample contains exactly two defectives is $P(AB)$. By application of the multiplication theorem, one obtains from Equation 3.28,

$$P(AB) = P(A)P(B \mid A)$$

$$P(AB) = \left(\frac{5}{100}\right)\left(\frac{4}{99}\right) = 0.002$$

Illustrative Example 3.3

Suppose that 50% of a company's manufactured output comes from a New York plant, 30% from a Pennsylvania plant, and 20% from a Delaware plant. On the basis of plant records, it is estimated that defective items constitute 1% of the output of the New York plant, 3% of the Pennsylvania plant, and 4% of the Delaware plant. If an item selected at random from the company's manufactured output is found to be defective, what are the revised probabilities that the item was produced by each of the three plants?

Solution

Let A_1, A_2, A_3 denote, respectively, the events that the item was produced in the New York, Pennsylvania, and Delaware plants. Let B denote the event that the item was found to be defective. Then,

$$P(A_1) = 0.50; \quad P(B \mid A_1) = 0.01$$

$$P(A_2) = 0.30; \quad P(B \mid A_2) = 0.03$$

$$P(A_3) = 0.20; \quad P(B \mid A_3) = 0.04$$

Apply Bayes' theorem. Substituting in Equation 3.36 leads to

$$P(A_1 \mid B) = \frac{P(A_i)P(B \mid A_i)}{P(A_1)P(B \mid A_1) + P(A_2)P(B \mid A_2) + P(A_3)P(B \mid A_3)}$$

$$P(A_1 \mid B) = \frac{(0.50)(0.01)}{(0.50)(0.01) + (0.30)(0.03) + (0.20)(0.04)} = 0.23$$

Similarly,

$$P(A_2 \mid B) = 0.41; \quad P(A_3 \mid B) = 0.36$$

Therefore, the information that the item selected at random was defective revises the probability that the item was produced in the New York plant downward from 0.50 to 0.23 and the probabilities for Pennsylvania and Delaware upward, respectively, from 0.30 to 0.41 and from 0.20 to 0.36.

Illustrative Example 3.4

The average weekly wastewater discharge temperature (°C) for 6 consecutive weeks are

$$22, 10, 8, 15, 13, 18$$

Find the median, the arithmetic mean, the geometric mean, and the standard deviation.

Solution

To obtain the median, the data provided must first be arranged in order of magnitude, such as:

$$8, 10, 13, 15, 18, 22$$

Thus, the median is 14, or the value halfway between 13 and 15, since this data set has an even number of measurements.

The geometric mean can be calculated using Equation 3.38

$$\overline{X}_G = \sqrt[n]{(X_1)(X_2)\cdots(X_n)} \tag{3.38}$$

For the given wastewater temperatures (substituting T for X),

$$\overline{T}_G = \left[(8)(10)(13)(15)(18)(22) \right]^{1/6} = 13.54°C$$

while the arithmetic mean, $\overline{T}$, is

$$\overline{T} = (8 + 10 + 13 + 15 + 18 + 22)/6$$
$$= 14.33°C$$

For the *standard deviation*, the defining formula is given by the expression:

$$\sigma = \sqrt{\frac{\sum\left(X_i - \bar{X}\right)^2}{n-1}} \qquad (3.39)$$

or (when applied to the temperature, T)

$$\sigma = \sqrt{\frac{\sum\left(T_i - \bar{T}\right)^2}{n-1}} = \sqrt{\frac{n\sum\left(T_i^2 - \left(\sum T_i\right)^2\right)}{n(n-1)}} \qquad (3.40)$$

The standard deviation may be calculated for the data at hand:

$$\sum T_i^2 = (8)^2 + (10)^2 + (13)^2 + (15)^2 + (18)^2 + (22)^2 = 1366$$
$$\left(\sum T_i\right)^2 = (8 + 10 + 13 + 15 + 18 + 22)^2 = 7396$$

Thus,

$$\sigma = \sqrt{\frac{6(1366) - 7396}{(6)(5)}} = 5.16°C$$

and

$$\sigma^2 = 31.8°C$$

REFERENCES

1. *Webster's College Dictionary*, 2nd edn., Prentice Hall, Upper Saddle River, NJ, 1971.
2. B. Carnahan, H. Luther, and J. Wilkes, *Applied Numerical Methods*, John Wiley & Sons, Hoboken, NJ, 1969.
3. L. Theodore and F. Taylor, *Probability and Statistics*, Theodore Tutorials (originally published by USEPA, RTP, NC), East Williston, NY, 1996.
4. F. Lees, *Loss Prevention in the Process Industry*, Vol. 1, Butterworth, Boston, MA, 1980.

4 Introduction to Probability Distributions

INTRODUCTION

The probability distribution of a random variable concerns the distribution of probability over the range of the random variable. The distribution of probability is specified by the *probability distribution function* (pdf). This chapter is devoted to providing general properties of the pdf for the case of *discrete* and *continuous* random variables. These pdfs find extensive application in environmental risk analysis.

One way to express a likelihood quantitatively is to use a numerical value, termed the probability, to express its likelihood of occurrence. The statement that there is a 2% chance of an accident occurring is obviously more precise and less vague than saying the chance of an accident is very low. The probability can be expressed as a fractional number, for example, 0.37, or a percent number from 0% to 100%, for example, 37%. Naturally, the sum of fractional probabilities for all possible states of occurrence must be 1.0.

The probability variation is another factor that needs to be considered. This includes not only the variations of the reported single-valued data but also probability variations with time, for example, the chance of contracting cancer or the annual probability of an earthquake occurring of a given magnitude or the probability variation with time of a NASA spacecraft falling immediately after liftoff. It is for this reason that this chapter treats the general subject of probability distributions, particularly as it applies to health problems and hazards.

Before proceeding to probability distributions, it behooves the reader to grasp the concept of the pdf. In mathematics, a *function* is defined as a relationship between a quantity that depends on another quantity or quantities. *Probability distributions* are an integral part of the general subject of statistics. There are three distributions of concern that arise in risk analysis studies:

1. *Probability distribution function (pdf)*. A pdf is a distribution of probabilities of the values of a dependent variable as a function of a value of an independent variable (often time). In the context of health risk assessment, the pdf represents the probability that a given health problem will occur *at* or *before* a specified time.
2. *Probability density function*. The probability density function describes the relative values of the probability (or likelihood of the occurrence) of *all* possible values of the independent variable.
3. *Cumulative distribution function (cdf)*. A cdf is the cumulative sum of all probabilities of a dependent variable *less than* or *equal* to a specific value

of an independent variable (often time). In hazard risk assessment, a cdf provides information on ascending (or increasing) values of the accident probabilities at increasing values of operating time or time since a previous component or system failure.

This chapter initially examines various elementary forms of data representation. Specific discrete pdfs reviewed in Section II include the binomial, multinomial, hypergeometric, and Poisson distributions. Other discrete distributions receive peripheral treatment. Specific continuous pdfs are reviewed in Section III following the treatment of discrete pdfs include the normal, log-normal, experimental, and Weibull distributions. Other continuous distributions also receive treatment.

The remaining sections of this chapter address the following five topics:

1. Data Representation
2. Discrete Probability Distributions
3. Continuous Probability Distributions
4. Expected Values
5. Illustrative Examples

Much of the material presented in this chapter has been adopted from the literature.[1,2]

DATA REPRESENTATION

The presentation in this section is based on what one would describe as an illustrative example. Consider the following set of data (see Table 4.1), which represents polychlorinated biphenyl (PCB) levels in a contaminated water stream at a given time for 25 days. As the first step in summarizing the data, you are requested to form a frequency table, a frequency polygon, a cumulative frequency table, and a cumulative frequency distribution curve.

Data are usually unmanageable in the form in which they are collected. In this section, the graphical techniques of summarizing such data so that meaningful information can be extracted from them are considered. Basically, there are two kinds of variables to which data can be assigned: continuous variables and discrete variables. A continuous variable is one that can assume any value in some interval of values. Examples of continuous variables are weight, volume, length, time, and temperature. Most environmental data are taken from continuous variables. Discrete variables, on the other hand, are those variables for which the possible values are integers. Therefore, they involve counting rather than measuring. Examples of discrete variables are the number of sample stations, number of people in a room, and number of times an environmental control standard is violated.

Since any measuring device is of limited accuracy, measurements in real life are actually discrete in nature rather than continuous, this should not keep one from regarding such variables as continuous. When a weight is recorded as 165 lb, it is assumed that the actual weight is somewhere between 164.5 and 165.5 lb. (See several examples in Section IV.)

A frequency table of the data in Table 4.1 is first constructed. See Table 4.2.

TABLE 4.1
PCB Concentration as a Function of Time

Days	PCB Concentration (ppb)
1	53
2	72
3	59
4	45
5	44
6	85
7	77
8	56
9	157
10	83
11	120
12	81
13	35
14	63
15	48
16	180
17	94
18	110
19	51
20	47
21	55
22	43
23	28
24	38
25	26

TABLE 4.2
PCB Frequency Table

Class Interval (ppb)	Frequency of Occurrence
25–40	4
40–55	7
55–70	4
70–85	4
85–100	2
100–115	1
115–130	1
130–145	0
145–160	1

TABLE 4.3

PCB Cumulative Frequency Table

PCB Level	Cumulative Frequency	Fractional Cumulative Frequency of Concentrations Less Than Stated Value
≤40	4	0.16
≤55	11	0.44
≤70	15	0.60
≤85	19	0.76
≤100	21	0.84
≤115	22	0.88
≤130	23	0.92
≤145	23	0.92
≤160	24	0.96
≤175	24	0.96
≤190	25	1.00

In constructing a cumulative frequency table (shown in Table 4.3), it can be seen that the data have been divided into 11 class intervals with each interval being 15 units in length. The choice of dividing the data into 11 intervals was purely arbitrary. However, in dealing with data, it is a rule of thumb to choose the length of the class interval such that 8–15 intervals which will include all of the data under consideration. Deriving the frequency column involves nothing more than counting the number of values in each interval. From the observation of the frequency table, one can now see that recorded data taking form. The values appear to be clustered between 25 and 85 ppb. In fact, nearly 80% are in this interval.

As a further step, one can graph the information in a frequency table. One way of doing this would be to plot the frequency midpoint of the class interval. The solid line connecting the points of Figure 4.1 forms a frequency polygon.

Another method of graphing the information would be by constructing a histogram as shown in Figure 4.2. The histogram is a two-dimensional graph in which the length of the class interval is taken into consideration. The histogram can be a very useful tool in statistics, especially if one converts the given frequency scale to a relative scale so that the sum of all the ordinates equals one. This is also shown in Figure 4.1. Thus, each ordinate value is derived by dividing the original value by the number of observations in the sample, in this case 25. The advantage in constructing a histogram like this is that one can read probabilities from it if one can assume a scale on the abscissa such that a given value will fall in any one interval in the area under the curve in that interval. For example, the probability that a value will fall between 55 and 70 is equal to its associated interval's portion of the total area of intervals, which is 0.16.

One can also construct a cumulative frequency table (Table 4.3) and graph (Figure 4.3) from the frequency table and histogram discussed earlier.

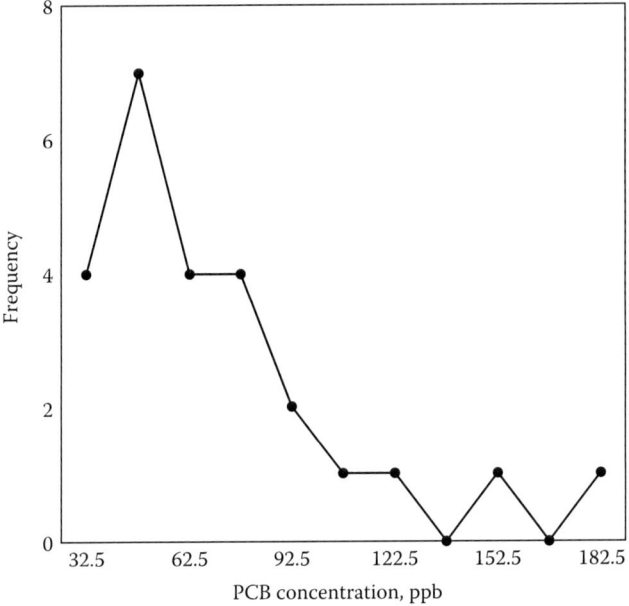

FIGURE 4.1   Pollution concentration (midpoint of class interval) frequency polygon.

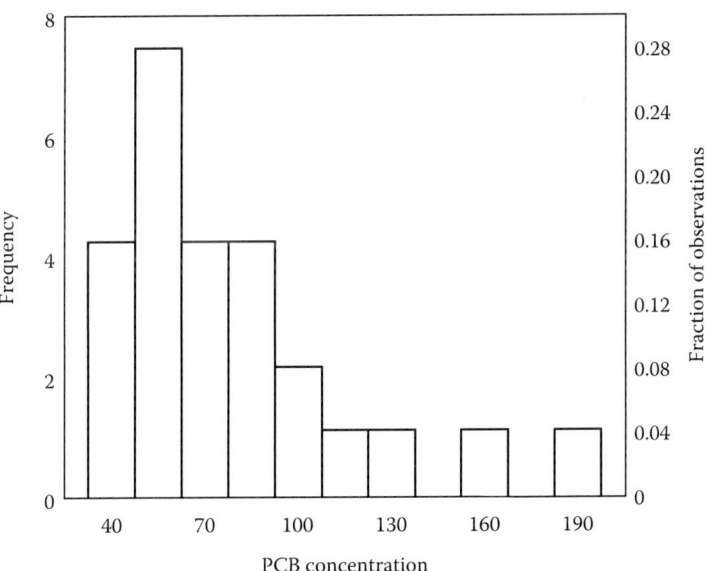

FIGURE 4.2   Histogram of percent frequency distribution curve.

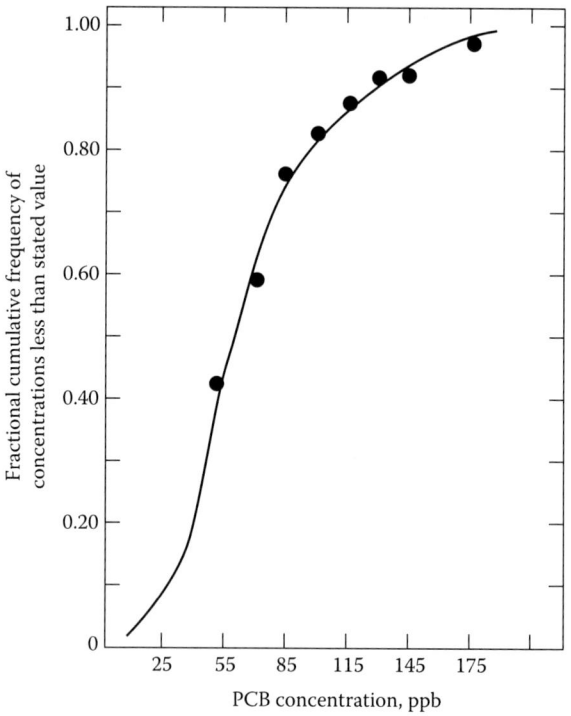

FIGURE 4.3 Cumulative frequency distribution curve.

The cumulative frequency table gives the number of observations less than a given value. Probabilities can be read from the cumulative frequency curve (Figure 4.3). For example, to find the probability that a value will be less than 85, one should read the curve at the point $x = 85$ and read across to the approximate value of 0.74 on the y axis.

DISCRETE PROBABILITY DISTRIBUTIONS

The pdf of a discrete random variable X is specified by $f(x)$ where $f(x)$ has the following essential properties:

1. $f(x) = P(X = x)$

      = probability assigned to the outcome corresponding

         to the number x in the range of X (4.1)

2. $f(x) \geq 0$ (4.2)

3. $\sum_x f(x) = 1$ (4.3)

Property 1 indicates that the pdf of a discrete random variable generates probability by substitution. Properties 2 and 3 restrict the values of $f(x)$ to nonnegative real numbers whose sum is 1.

Consider, for example, a box of 100 transistors containing five defectives. Suppose that a transistor selected at random is to be classified as defective or nondefective. Let X denote the outcome, with $X=0$ associated with the drawing of a nondefective and $X=1$ associated with the drawing of a defective. Then, X is a discrete random variable with pdf specified by

$$f(x) = 0.05; \quad x = 1$$
$$= 0.95; \quad x = 0$$

For another example of the pdf of a discrete random variable, let X denote the number of throws on which the first failure of an electrical switch occurs. Suppose that the probability that a switch fails on any throw is 0.001 and that successive throws are independent with respect to failure. If the switch fails for the first time on throw x, it must have been successful on each of the preceding $x - 1$ trials. Therefore, the pdf of X is given by

$$f(x) = (0.999)^{x-1}(0.001); \quad x = 1, 2, 3, \ldots, n, \ldots$$

Note that the range of X consists of a countable infinitude of values. Verification of the earlier property 3 for pdfs of discrete random variables can be accomplished by noting that $\sum_x f(x)$ is a geometric series with the first term equal to 0.001, a common ratio equal to 0.999, and therefore convergent to $(0.001)/(1-0.999)$, which is 1, that is, unity.

Another function used to describe the probability distribution of a discrete random variable X is the *cdf*. If $f(x)$ specifies the pdf of a random variable X, then $F(x)$ is used to specify the cdf. For both discrete and continuous random variables, the cdf of X is defined by

$$F(x) = P(X \le x); \quad -\infty < x < \infty \tag{4.4}$$

Note that the cdf is defined for all real numbers, not just the values assumed by the random variable.

To illustrate the derivation of the cdf from the pdf, consider the case of a discrete random variable X whose pdf is specified by

$$f(x) = 0.2; \quad x = 2$$
$$= 0.3; \quad x = 5$$
$$= 0.5; \quad x = 7$$

Applying the definition of cdf in Equation 4.4, one obtains for the cdf of X

$$F(x) = 0; \quad x < 2$$
$$= 0.2; \quad 2 \le x < 5$$
$$= 0.5; \quad 5 \le x < 7$$
$$= 1; \quad x \ge 7$$

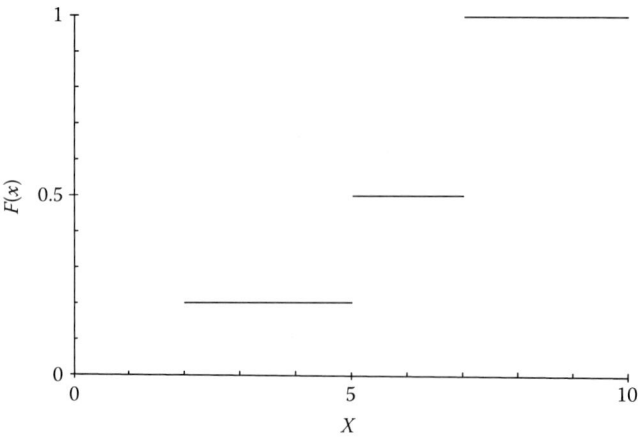

FIGURE 4.4 Graph of the cdf of a discrete random variable X.

It is helpful to think of $F(x)$ as an accumulator of probability as x increases through all real numbers. In the case of a discrete random variable, the cdf is a step function increasing by finite jumps at the values of x in the range of X. In the earlier example, these jumps occur at the values 2, 5, and 7. The magnitude of each jump is equal to the probability assigned to the value where the jump occurs. This is depicted in Figure 4.4.

The following properties of the cdf of a discrete random variable X can be deduced directly from the definition of $F(x)$ in Equation 4.4.

1. $F(b) - F(a) = P(a < X \le b)$ (4.5)
2. $F(+\infty) = 1$ (4.6)
3. $F(-\infty) = 0$ (4.7)
4. $F(x)$ is a nondecreasing function of x (4.8)

As will be noted in the next section, these properties also apply to continuous random variables.

CONTINUOUS PROBABILITY DISTRIBUTIONS

The pdf of a continuous random variable X has the following properties:

1. $\displaystyle\int_a^b f(x)\,dx = P(a < X < b)$ (4.9)

2. $f(x) \ge 0$ (4.10)

3. $\displaystyle\int_{-\infty}^{+\infty} f(x)\,dx = 1$ (4.11)

Property 1 indicates that the pdf of a continuous random variable generates probability by the integration of the pdf over the interval whose probability is required. When this interval contracts to a single value, the integral over the interval becomes zero. Therefore, the probability associated with any particular value of a continuous random variable is zero. Consequently, if X is continuous,

$$\begin{aligned} P(a \leq X \leq b) &= P(a < X \leq b) \\ &= P(a < X < b) \\ &= P(a \leq X < b) \end{aligned} \qquad (4.12)$$

Property 2 restricts the values of $f(x)$ to nonnegative numbers. Property 3 follows from the fact that

$$P(-\infty \leq X \leq \infty) = 1 \qquad (4.13)$$

As an example of the pdf of a continuous random variable, consider the pdf of the time X, in hours, between successive failures of an aircraft air-conditioning system. Suppose the pdf of X is specified by

$$\begin{aligned} f(x) &= 0.01\, e^{-0.01x}; && x > 0 \\ &= 0; && \text{elsewhere} \end{aligned}$$

Inspection of the graph in Figure 4.5 indicates that intervals in the lower part of the range of X are assigned greater probabilities than intervals of the same length in the upper part of the range of X because the areas over the former are greater than

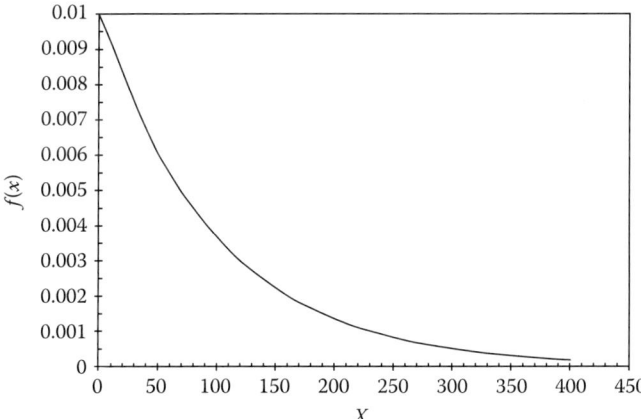

FIGURE 4.5 The pdf of time, in hours, between successive failures of an aircraft air-conditioning system.

the areas over the latter. The expression $P(a < X < b)$ can be interpreted geometrically as the area under the pdf curve over the interval (a, b). Integration of the pdf over the interval yields the probability assigned to the interval. For example, the probability that the time in hours between successive failures of the aforementioned aircraft air-conditioning system is greater than 6 but less than 10 is

$$P(6 < X < 10) = \int_{6}^{10} 0.01 e^{-0.01x} dx$$

$$= 0.04 = 4.0\%$$

Another function used to describe the probability distribution of a continuous random variable X is the *cdf*. If $f(x)$ specifies the pdf of a random variable X, then $F(x)$ is used to specify the cdf. For both discrete and continuous random variables, the cdf of X is defined by

$$F(x) = P(X \leq x); \quad -\infty < x < \infty \tag{4.14}$$

Note once again that the cdf is defined for all real numbers, not just the values assumed by the random variable.

To illustrate the derivation of the cdf from the pdf, consider the case of a continuous random variable. The cdf is a continuous function. Suppose, for example, that X is a continuous random variable with pdf specified by

$$f(x) = 2x; \quad 0 \leq x < 1$$
$$= 0; \quad \text{elsewhere} \tag{4.15}$$

Applying the definition of the cdf in Equation 4.14, one obtains

$$F(x) = 0; \quad\quad\quad\quad x < 0$$
$$= \int_{0}^{x} 2x\, dx = x^2; \quad 0 \leq x < 1$$
$$= 1; \quad\quad\quad\quad x \geq 1 \tag{4.16}$$

Figure 4.6 displays the graph of this cdf.

The pdf of a continuous random variable can also be obtained by differentiating its cdf and setting the pdf equal to zero where the derivative of the cdf does not exist. For example, differentiating the cdf obtained in Equation 4.16 yields the pdf in Equation 4.15. In this case, the derivative of the cdf does not exist for $x = 1$.

The following properties of the cdf of a continuous random variable X can be deduced directly from the definition of $F(x)$ in Equation 4.14.

1. $F(b) - F(a) = P(a < X \leq b)$ \hfill (4.17)

2. $F(+\infty) = 1$ \hfill (4.18)

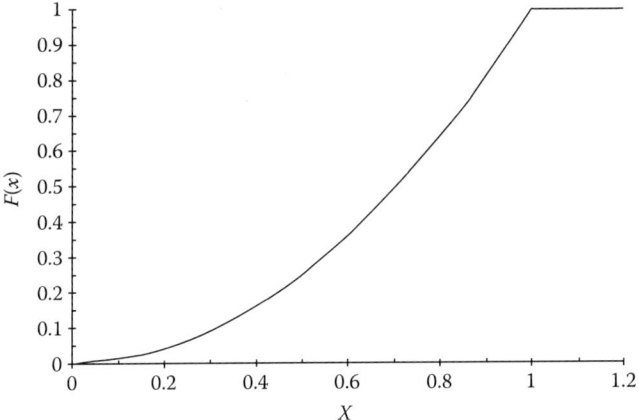

FIGURE 4.6 Graph of the cdf of a continuous random variable, X.

3. $F(-\infty) = 0$ (4.19)

4. $F(x)$ is a nondecreasing function of x (4.20)

Note that these properties also apply to the discrete random variables discussed in the previous section.

EXPECTED VALUES

The expected value of a random variable is defined as the average value of the random variable. The expected value of a random variable X is denoted by $E(X)$. The expected value of a random variable can be interpreted as the long-run average of observations on the random variable. The procedure for calculating the expected value of a random variable depends on whether the random variable is discrete or continuous.

If X is a discrete random variable with pdf specified by $f(x)$, then

$$E(X) = \sum_x xf(x)$$ (4.21)

If X is a continuous random variable with pdf specified by $f(x)$, then

$$E(X) = \int_{-\infty}^{+\infty} xf(x)\, dx$$ (4.22)

Suppose, for example, that the pdf of a *discrete* random variable X is specified by

$$
\begin{aligned}
f(x) &= 0.3; & x &= 10 \\
&= 0.2; & x &= 20 \\
&= 0.5; & x &= 30
\end{aligned}
$$

Then, the expected value of X is given by

$$E(X) = 10(0.3) + 20(0.2) + 30(0.5)$$
$$= 22$$

Suppose, for example, that the pdf of a *continuous* random variable x is specified by

$$f(x) = e^x; \quad x > 0$$
$$= 0; \quad \text{elsewhere}$$

Application of Equation 4.22 yields the following expected value for X.

$$E(X) = \int_0^\infty x e^{-x} dx = 1$$

The expected value of a random variable X is also called the *mean* of X and is often designated by μ. The expected value of $(X - \mu)^2$ is called the *variance* of X. The positive square root of the variance is called the *standard deviation*. The terms σ^2 and σ (sigma squared and sigma) represent the variance and standard deviation, respectively. As noted earlier, variance is a measure of the spread or dispersion of the values of the random variable about its mean value. The standard deviation is also a measure of spread or dispersion. The standard deviation is expressed in the same units as X, while the variance is expressed in the square of these units.

The variance of x can be calculated directly from the definition given above

$$\sigma^2 = E(X - \mu)^2 \tag{4.23}$$

However, it is usually calculated more easily by the equivalent formula

$$\sigma^2 = E(X^2) - \mu^2 = E(X^2) - [E(X)]^2 \tag{4.24}$$

To illustrate the computation of variance and its interpretation in the case of *discrete* random variables, consider a random variable X having pdf specified by

$$f(x) = 0.5; \quad x = 1$$
$$= 0.5; \quad x = -1$$

and a random variable Y having pdf specified by

$$g(y) = 0.5; \quad y = 10$$
$$= 0.5; \quad y = -10$$

It is easily verified that both X and Y have the same expected value: zero. From Equation 4.21, the expected value of X^2 is given by

$$E(X^2) = \sum_x x^2 f(x)$$
$$= (1)(0.5) + (-1)^2(0.5)$$
$$= 1$$

Therefore, the variance of X is 1. The expected value of Y^2 is given by

$$E(Y^2) = \sum_y y^2 g(y)$$
$$= (10)^2(0.5) + (-10)^2(0.5)$$
$$= 100$$

Therefore, the variance of Y is 100. The standard deviation of X is 1, and the standard deviation of Y is 10. The larger value for the standard deviation in the case of Y reflects the greater dispersion of Y values about their mean.

To illustrate the computation of variance and its interpretation in the case of *continuous* random variables, consider a random variable X having pdf specified by

$$f(x) = 1/2; \quad 0 < x < 2$$
$$= 0; \quad \text{elsewhere}$$

and a random variable Y having pdf specified by

$$g(y) = 1/4; \quad -1 < y < 3$$
$$= 0; \quad \text{elsewhere}$$

The mean value of X from Equation 4.22 is given by

$$E(X) = \int_0^2 x \left(\frac{1}{2}\right) dx$$
$$= 1$$

The expected value of X^2 is given by

$$E(X^2) = \int_0^2 x^2 \left(\frac{1}{2}\right) dx$$
$$= 4/3$$

Therefore, the variance of X from Equation 4.23 is

$$\sigma^2 = E(X^2) - \mu^2$$
$$= 4/3 - 1 = 1/3$$

The mean value of Y is given by

$$E(Y) = \int_{-1}^{3} y \left(\frac{1}{4} \right) dy$$
$$= 1$$

The expected value of Y^2 is given by

$$E(Y^2) = \int_{-1}^{3} y^2 \left(\frac{1}{4} \right) dy$$
$$= 7/3$$

Therefore, the variance of Y is

$$\sigma^2 = E(Y^2) - \mu^2$$
$$= 7/3 - 1 = 4/3$$

Here, X and Y have the same expected values, but Y has the larger variance once again, reflecting the greater dispersion of Y values about their mean.

ILLUSTRATIVE EXAMPLES

Illustrative Example 4.1

Discuss the problems associated with the need to obtain representative samples for statistical analysis.

Solution

One of the main problems with sampling is the need to obtain samples that are representative of the population. This usually requires both taking a large enough number of samples and doing it randomly. To sample randomly is to select from a population such that each sample drawn has an equal chance of being chosen. This requires including the whole population when selecting and removing all possible bias(es) from the selection process. It also requires sampling from the population in such a manner that each removal of a sample is taken into account on a subsequent sample analysis.

As an example of what can happen in sampling, consider this notorious failure: A magazine took a telephone survey of eligible voters and on the basis of this survey incorrectly predicted that Landon would beat Roosevelt in the 1936 U.S.

presidential election. The problem: The sample included only those voters who had telephones and was not representative because voters without telephones have no chance of being included. An analogy of this sampling error in a process application would be to draw samples only during the weekday shift and ignore the weekend period entirely.

It is also important to see how a sample compares with the overall population in terms of a measurable characteristic. If the sample matches the population on this characteristic, it is often valid to conclude that it matches the population on others. However, this assumption is not always valid, so that it is a weakness in the concept of sampling.

There are so many kinds of populations and samples in environmental risk applications that it may not be possible to guarantee a representative sample by any single method. The reader is cautioned on this matter, and it is recommended that care be exercised.

Illustrative Example 4.2

The pdf of a discrete random variable X is specified by

$$f(x) = 0.2; \quad x = 8$$
$$= 0.5; \quad x = 10$$
$$= 0.3; \quad x = 12$$

Calculate the expected value of X.

Solution
The expected value of X is given by

$$E(X) = 8(0.2) + 10(0.5) + 12(0.3)$$
$$= 1.6 + 5.0 + 3.6$$
$$= 10.2$$

Illustrative Example 4.3

Estimate the unknown population mean μ on the basis of the following random *sample* numerical values:

1
6
2
4
7

Solution
The sample mean is a random variable in the sense that its value varies from sample to sample from the same population. Substitute into

$$\bar{X} = \sum_{i=1}^{n} \frac{x_i}{n}$$

For the given sample, $n=5$ and $\sum x_i = 20$. Therefore,

$$\overline{X} = \frac{20}{5}$$
$$= 4$$

Illustrative Example 4.4

Refer to the previous example. *Estimate* the population variance.

Solution

The sample variance, like the population variance, is a measure of dispersion. The sample variance measures the dispersion of the sample observations about their mean. Like the sample mean, the sample variance is a random variable in the sense that it varies from sample to sample from the same population. The sample variance s^2 is defined as follows:

$$s^2 = \frac{\sum (X_i - \overline{X})^2}{n-1}$$

For the given sample, $\overline{X} = 4$ and $n=5$. Therefore,

$$s^2 = [(1-4)^2 + (6-4)^2 + (2-4)^2 + (4-4)^2 + (7-4)^2]/4$$
$$= 26/4$$
$$= 6.5$$

An alternate formula for s^2 is given by

$$s^2 = \frac{\sum X_i^2}{n-1} - \frac{\left(\sum X_i\right)^2}{n(n-1)}$$

For the given sample, $\sum X_i^2 = 106$, $\sum X_i = 20$, and $n=5$. Therefore,

$$s^2 = 106/4 - (20)^2/[5(4)]$$
$$= 26.5 - 20$$
$$= 6.5$$

Illustrative Example 4.5

A *discrete* random variable X assumes the values 49, 50, and 51, each with a probability 1/3. A discrete random variable Y assumes the values 0, 50, and 100 each with probability 1/3. Compare the means and variances of X and Y.

Solution

The pdf of a discrete random variable provides information on the probability assigned to each of the possible values of the random variable. Use the formulas (from Equation 4.21)

$$\mu_x = E(X) = \sum x\, f(x)$$
$$\mu_y = E(Y) = \sum y\, g(x)$$

to obtain the mean of X and Y, respectively.

$$\mu_x = 49(1/3) + 50(1/3) + 51(1/3)$$
$$= 50$$
$$\mu_y = 0(1/3) + 50(1/3) + 100(1/3)$$
$$= 50$$

Use the formulas (from Equation 4.23)

$$\sigma_x^2 = E(X - \mu_x)^2$$
$$\sigma_y^2 = E(Y - \mu_y)^2$$

to obtain the variance of X, Y, respectively. Substitution yields

$$\sigma_x^2 = \sum (x - \mu_x)^2 f(x)$$
$$= \sum (x - 50)^2 f(x)$$
$$= (49 - 50)^2(1/3) + (50 - 50)^2(1/3) + (51 - 50)^2(1/3)$$
$$= 2/3$$
$$\sigma_y^2 = \sum (y - \mu_y)^2 g(y)$$
$$= \sum (y - 50)^2 g(y)$$
$$= (0 - 50)^2(1/3) + (50 - 50)^2(1/3) + (100 - 50)^2(1/3)$$
$$= 5000/3$$

Illustrative Example 4.6

Refer to the previous illustrative example. Compare the means and variances of X and Y and interpret the results.

Solution

Recall that the variance measures the spread or dispersion about the mean. X and Y have the same mean, but the variances of Y is greater than the variances of X, which reflects the greater dispersion (spread, variability) of the values of Y about its mean in contrast to the dispersion of the values of X about its mean.

Illustrative Example 4.7

Continuous random variables X and Y have pdfs specified by $f(x)$ and $g(y)$, respectively, as follows:

$$f(x) = 1/2; \quad -1 < x < 1$$
$$g(y) = 1/4; \quad -2 < y < 2$$

Compute the mean and variance of X and Y.

Solution

Note that

$$\mu_x = \int_{-\infty}^{+\infty} xf(x)\,dx$$

$$\mu_y = \int_{-\infty}^{+\infty} yg(y)\,dy$$

Compute μ_x, the mean of X; and μ_y, the mean of Y using the data provided.

$$\mu_x = \int_{-1}^{1} x(1/2)\,dx = 0$$

$$\mu_y = \int_{-2}^{2} y(1/4)\,dy = 0$$

Compute σ_x^2 and σ_y^2, the variances of X and Y, respectively, using the following equations:

$$\sigma_x^2 = E(X^2) - \mu_x^2$$
$$= \int_{-1}^{1} x^2 1/2\,dx$$
$$= 1/3$$
$$\sigma_y^2 = E(Y^2) - \mu_y^2$$
$$= \int_{-2}^{2} y^2 1/4\,dy$$
$$= 4/3$$

Illustrative Example 4.8

Interpret the results of the previous illustrative example.

Solution

Once again, X and Y have the same mean, 0. The variance of Y is greater than the variance of X, which reflects the greater dispersion of the values of Y about its mean.

Illustrative Example 4.9

A continuous random variable X has a pdf of $x/2$ for $0 \leq x \leq 2$. What is the cdf?

Solution

By definition, for continuous variables,

$$F(x) = \int f(x)\,dx$$

Substituting

$$F(x) = \int \frac{x}{2}\,dx$$

$$= \frac{x^2}{4}; \quad 0 \leq x \leq 2$$

Applying the definition of cdf leads to

$$F(x) = 0; \qquad x < 0$$

$$F(x) = \frac{x^2}{4}; \quad 0 \leq x < 2$$

$$F(x) = 1; \qquad x \geq 2$$

REFERENCES

1. L. Theodore and F. Taylor, *Probability and Statistics*, Theodore Tutorials (originally published by USEPA, RTP, NC), Theodore Tutorials, East Williston, NY, 1996.
2. S. Shaefer and L. Theodore, *Probability and Statistics Applications in Environmental Science*, CRC Press/Taylor & Francis Group, Boca Raton, FL, 2007.

Section II

Discrete Probability Distributions

Satires, Epistles, and Odes of Horace
Satire's my weapon, but I'm too discreet
To run amuck, and tilt at all I meet.

Alexander Pope (1733–1738)

This section contains four chapters.

Chapter 5: Binomial Distribution
Chapter 6: Multinomial Distribution
Chapter 7: Hypergeometric Distribution
Chapter 8: Poisson Distribution

The four chapters are concerned with *discrete* probability distributions. It was decided not to include a fifth chapter on "other distributions." Finally, *continuous probability distributions* receive treatment in Section III.

5 Binomial Distribution

INTRODUCTION

Consider n independent performances of a random experiment with mutually exclusive outcomes that can be classified as *success* or *failure*. The words success and failure are to be regarded as labels for two mutually exclusive categories of outcomes of the random experiment. Yet, they may not necessarily have the ordinary connotation of success or failure. Assume that p, the probability of success on any performance of the random experiment, is constant. Let q be the probability of failure, so that

$$q = 1 - p \tag{5.1}$$

The probability distribution of X, the number of successes in n performances of the random experiment, is the binomial distribution, with a probability distribution function (pdf) specified by

$$f(x) = \frac{n!}{x!(n-x)!} p^x q^{n-x}, \qquad x = 0, 1, \ldots, n \tag{5.2}$$

where $f(x)$ is the probability of x successes in n performances. One can show that the expected value of a random variable X is np and its variance is npq. For example, if the probability of defective thermometer is 0.1, one can determine (1) the mean and (2) the standard deviation for the distribution of defective thermometers in a total of 400. For this case, the expected value or mean is $(400)(0.1) = 40$, that is, one can therefore *expect* 40 thermometers to be defective. The variance is $npq = (400)(0.1)(0.9) = 36$; thus, the standard deviation $= \sqrt{36} = 6$.[1,2]

As a simple example of the binomial distribution, consider the probability distribution of the number of defectives in a sample of 5 items drawn *with* replacement from a lot of 1000 items, 50 of which are defective. Associate success with drawing a defective item from the lot. Then the result of each drawing can be classified as success (defective item) or failure (nondefective item). The sample of five items is drawn with replacement (i.e., each item in the sample is returned before the next is drawn from the lot; therefore, the probability of success remains constant at 0.05). Substituting in Equation 5.2 the values $n = 5$, $p = 0.05$, and $q = 0.95$ yields

$$f(x) = \frac{5!}{x!(5-x)!} (0.05)^x (0.95)^{5-x}, \qquad x = 0, 1, 2, 3, 4, 5$$

as the pdf for X, the number of defectives in the sample. The probability that the sample contains exactly three defectives is therefore given by

$$P(X=3)=\frac{5!}{(3!)(2!)}(0.05)^3(0.95)^2$$
$$=0.0011=0.011\%$$

The author recently bet on 10 basketball games at the Mirage simulcasting center in Las Vegas. Assuming the odds of winning the bet are 0.5, what is the probability of breaking even, that is, winning 5 of the bets. For this application, Equation 5.2 applies with $n=10$, $p=0.5$, and $x=5$.

$$P(X=5)=\frac{10!}{(5!)(5!)}(0.5)^5(0.5)^5$$
$$=(252)(0.03125)(0.03125)$$
$$=0.246=24.6\%$$

Consider also how many games the author would have to bet so that the probability of breaking even is

1. 50%
2. 25%

For case 1, the number of games is given by

$$P(X=n/2)=\frac{2!}{(1!)(1!)}(0.5)^n$$
$$0.5=(2)(0.5)^n$$
$$n=2 \text{ (as expected)}$$

For case 2, no exact integer number of games can be determined

$$P(X=n/2)=(252)(0.5)^n$$
$$0.25=(252)(0.5)^n$$
$$n=9.98 \text{ (by trial and error)}$$

or alternatively,

$$\frac{0.25}{252} = (0.5)^n$$

$$\log\left(\frac{0.25}{252}\right) = n\log(0.5)$$

$$n = \frac{\log\left(\dfrac{0.25}{252}\right)}{\log(0.5)}$$

$$= 9.98$$

This number when rounded off to the nearest integer is 10, a result that could have been deduced immediately from the previous example.

Note that the term

$$\frac{n!}{x!(n-x)!} \tag{5.3}$$

in Equation 5.2 is often denoted as

$$\binom{n}{x} \tag{5.4}$$

where

$$\binom{n}{x} = \frac{(n)(n-1)(n-2)\cdots(n-x+1)}{(x)(x-1)(x-2)\cdots3,2,1}$$

and

$$\binom{n}{0} = 1, \quad \binom{n}{1} = n$$

Also note that

$$\binom{n}{x} = \frac{n}{n-x} \tag{5.5}$$

Interestingly, a recurrence relation is available for the binomial coefficients. A *recurrence relation* is a rule that defines each element of a sequence in terms of the preceding elements. Recurrence relations are particularly useful for computing the elements of the sequence. The binomial coefficients satisfy the following recurrence relation:

$$\binom{n}{x+1} = \left[\frac{n-x}{x+1}\right]\binom{n}{x}$$

(5.6)

Note also that Stirling's formula

$$n! = \sqrt{2\pi n}\, n^{n+1/2} n^{-n}$$

(5.7)

may be employed in evaluating the various terms of Equation 5.2 for cases of large n, if factorial tables (or the equivalent) are not available.

In any event, one could use the recurrence relation to compute $\binom{5}{x}$, $(x = 1, 2, 3)$. First note that for $n=5$, $x=1$,

$$\binom{5}{1} = \binom{5}{1}\binom{5}{0} = 5(1) = 5$$

For the following two cases,

$$n=5, x=2:$$

$$\binom{5}{2} = \frac{5-1}{1+1} \times \binom{5}{1}$$
$$= 10$$

$$n=5, x=3:$$

$$\binom{5}{3} = \frac{5-2}{2+1} \times \binom{5}{2}$$
$$= 10$$

The binomial distribution can be used to calculate the reliability of a *redundant system*. A redundant system consisting of n identical components is a system that fails only if more than r components fail. Familiar examples include single-usage equipment such as missile engines, short-life batteries, and flash bulbs, which are required to operate for one time period and are not reused. Once again, associate success with the failure of a component. Assume that the n components are independent

TABLE 5.1
Binomial Probabilities

X	Binomial $n=20, p=0.05$	Binomial $n=100, p=0.01$
0	0.3585	0.3660
1	0.3774	0.3697
2	0.1887	0.1849
3	0.0596	0.0610
4	0.0133	0.0149
5	0.0022	0.0029
6	0.0003	0.0005
≥ 7	0.0000	0.0001

with respect to failure, and that the reliability of each is $1 - p$. Then X, the number of failures, has the binomial pdf in Equation 5.2 and the reliability of the redundant system is

$$P(X \leq r) = \sum_{x=0}^{r} \frac{n!}{x!(n-x)!} p^x q^{n-x} \tag{5.8}$$

Tables that can assist binomial distribution calculations are available in the literature. Probabilities for two different binomial coefficients in Equation 5.2 are presented in Table 5.1. Information on the binomial coefficients for $n=1,\ldots,30$ and $x=1,\ldots,15$ is provided in Table 5.2. For $x>15$, employ the relationship in Equation 5.5.

ILLUSTRATIVE EXAMPLES

Illustrative Example 5.1

A sample of 5 transistors is drawn with replacement from a manufacturing lot which is 5% defective.[1,2] Once again, "with replacement" means that each transistor drawn is returned to the lot before the next is drawn. What is the probability that the number of defective transistors in the sample is exactly 2?

Solution
This random experiment consists of drawing a transistor at random with replacement from a lot. The random experiment is performed five times because a sample of 5 transistors is drawn with replacement from the lot. Therefore, $n=5$. Also note that the performances are independent because each transistor is replaced before the next is drawn. Therefore, the composition of the lot is exactly the same before each drawing.

For this problem, once again associate success with drawing a defective transistor and associate failure with drawing a nondefective transistor. Refer once again

TABLE 5.2

Binomial Coefficient Chart, $\binom{n}{x}$; $n = 1 \rightarrow 30$, $x = 1 \rightarrow 15$

$$\binom{n}{x} = \frac{n!}{x!(n-x)!} = \frac{n(n-1)\cdots(n-x+1)}{x!} = \binom{n}{n-x}, \quad 0! = 1$$

n	0	1	2	3	4	5	6	7	8	9
1	1	1								
2	1	2	1							
3	1	3	3	1						
4	1	4	6	4	1					
5	1	5	10	10	5	1				
6	1	6	15	20	15	6	1			
7	1	7	21	35	35	21	7	1		
8	1	8	28	56	70	56	28	8	1	
9	1	9	36	84	126	126	84	36	9	1
10	1	10	45	120	210	252	210	120	45	10
11	1	11	55	165	330	462	462	330	165	55
12	1	12	66	220	495	792	924	792	495	220
13	1	13	78	286	715	1,287	1,716	1,716	1,287	715
14	1	14	91	364	1,001	2,002	3,003	3,432	3,003	2,002
15	1	15	105	455	1,365	3,003	5,005	6,435	6,435	5,005
16	1	16	120	560	1,820	4,368	8,008	11,440	12,870	11,440
17	1	17	136	680	2,380	6,188	12,376	19,448	24,310	24,310
18	1	18	153	816	3,060	8,568	18,564	31,824	43,758	48,620

(Continued)

TABLE 5.2 (Continued)

Binomial Coefficient Chart, $\binom{n}{x}$; $n=1\to30$, $x=1\to15$

$$\binom{n}{x} = \frac{n!}{x!(n-x)!} = \frac{n(n-1)\cdots(n-x+1)}{x!} = \binom{n}{n-x}, \quad 0! = 1$$

n	0	1	2	3	4	5	6	7	8	9
19	1	19	171	969	3,876	11,628	27,132	50,388	75,582	92,378
20	1	20	190	1,140	4,845	15,504	38,760	77,520	125,970	167,960
21	1	21	210	1,330	5,985	20,349	54,264	116,280	203,490	293,930
22	1	22	231	1,540	7,315	26,334	74,613	170,544	319,770	497,420
23	1	23	253	1,771	8,855	33,649	100,947	245,157	490,314	817,190
24	1	24	276	2,024	10,626	42,504	134,596	346,104	735,471	1,307,504
25	1	25	300	2,300	12,650	53,130	177,100	480,700	1,081,575	2,042,975
26	1	26	325	2,600	14,950	65,780	230,230	657,800	1,562,275	3,124,550
27	1	27	351	2,925	17,550	80,730	296,010	888,030	2,220,075	4,686,825
28	1	28	378	3,276	20,475	98,280	376,740	1,184,040	3,108,105	6,906,900
29	1	29	406	3,654	23,751	118,755	475,020	1,560,780	4,292,145	10,015,005
30	1	30	435	4,060	27,405	142,506	593,775	2,035,800	5,852,925	14,307,150

x

(*Continued*)

TABLE 5.2 (Continued)

Binomial Coefficient Chart, $\binom{n}{x}$ **;** $n = 1 \rightarrow 30$, $x = 1 \rightarrow 15$

$$\binom{n}{x} = \frac{n!}{x!(n-x)!} = \frac{n(n-1)\cdots(n-x+1)}{x!} = \binom{n}{n-x}, \quad 0! = 1$$

n	10	11	12	13	14	15
10	1					
11	11	1				
12	66	12	1			
13	286	78	13	1		
14	1,001	364	91	14	1	
15	3,003	1,365	455	105	15	1
16	8,008	4,368	1,820	560	120	16
17	19,448	12,376	6,188	2,380	680	136
18	43,758	31,824	18,564	8,568	3,060	816

(Continued)

TABLE 5.2 (Continued)

Binomial Coefficient Chart, $\binom{n}{x}$; $n=1\to30$, $x=1\to15$

$$\binom{n}{x} = \frac{n!}{x!(n-x)!} = \frac{n(n-1)\cdots(n-x+1)}{x!} = \binom{n}{n-x}, \quad 0! = 1$$

x

n	10	11	12	13	14	15
19	92,378	75,582	50,388	27,132	11,628	3,876
20	184,756	167,960	125,970	77,520	38,760	15,504
21	352,716	352,716	293,930	203,490	116,280	54,264
22	646,646	705,432	646,646	497,420	319,770	170,544
23	1,144,066	1,352,078	1,352,078	1,144,066	817,190	490,314
24	1,961,256	2,496,144	2,704,156	2,496,144	1,961,256	1,307,504
25	3,268,760	4,457,400	5,200,300	5,200,300	4,457,400	3,268,760
26	5,311,735	776,160	9,647,700	10,400,600	9,657,700	7,726,160
27	8,436,285	13,037,895	17,383,860	20,058,300	20,059,300	17,383,860
28	13,123,110	21,474,180	30,421,755	37,442,160	40,116,600	37,442,160
29	20,030,010	34,597,290	51,895,935	67,863,915	77,558,760	77,558,760
30	30,045,015	54,627,300	86,493,225	119,759,850	145,422,675	155,117,520

Note: For $x > 15$, use the fact that $\binom{n}{x} = \frac{n}{n-x}$

to Equation 5.2. Because 5% of the lot is defective, $p=0.05$. Therefore, $q=0.95$. Substitute these values $(n, p,$ and $q)$ in the binomial pdf provided in Equation 5.2

$$f(x) = \frac{5!}{x!(5-x)!}(0.05)^x (0.95)^{5-x}, \quad x = 0,1,2,3,4,5$$

Substitute the appropriate value of X to obtain the required probabilities.

$$P(\text{exactly two defectives, i.e., } X = 2) = \frac{5!}{(2!)(3!)}(0.05)^2 (0.95)^3$$
$$= 0.214 = 21.4\%$$

See also the Introduction in this chapter for a similar example.

Illustrative Example 5.2

The probability that an exposure to a nanocarcinogen will be fatal is 0.80. Find the probability that at least 9 will die from a group of 15 workers.

Solution
For this case,

$$P(\text{at least 9 will die}) = P(X \geq 9), \quad p = 0.8, q = 0.2$$

Substitute into Equation 5.2

$$= \sum_{x=9}^{15} \frac{15!}{x!(15-x)!}(0.8)^x (0.2)^{15-x}$$
$$= 0.0430 + 0.1032 + 0.1876 + 0.2501 + 0.2309 + 0.1319 + 0.0352$$
$$= 0.982 = 98.2\%$$

Illustrative Example 5.3

Refer to the previous example. Find the probability that from four to eight will die.

Solution
This calculation can be performed by longhand or obtained directly from binomial tables similar to that provided in Table 5.2. For this application,

$$P(4 \leq X \leq 8) = 1.0 - P(X \geq 9) - P(0 \leq X \leq 4)$$

One notes almost immediately that

$$P(0 \leq X \leq 4) = 0$$

Therefore,

$$P(4 \le X \le 8) = 1.0 - 0.982 - 0.0$$
$$= 0.018 = 1.8\%$$

Illustrative Example 5.4

A plant manufactures filter bags in very large lots by a standard production process. A customer selects a sample of 20 bags at random from each production lot and rejects the complete lot if he or she finds 4 or more bad bags. If, in fact, the production process yields exactly 10% defectives, what is the probability of lot rejection?

Solution

The problem may be rephrased as follows: "Given a binomial process with probability of success 0.1, what is the probability of 4 or more successes in 20 independent trials?" This equals

$$P(\text{lot rejection}) = \sum_{x=4}^{20} \frac{20!}{x!(20-x)!} (0.1)^x (0.9)^{20-x}$$

$$= 1 - \sum_{x=0}^{3} \frac{20!}{x!(20-x)!} 0.1^x \, 0.9^{20-x}$$

$$= 1 - (0.1216 + 0.2702 + 0.2852 + 0.1901)$$

$$= 1 - 0.8671$$

$$= 0.1329 = 13.3\%$$

Once again, this calculation can be performed by longhand or obtained directly from binomial tables.

Illustrative Example 5.5

A chemical reaction unit contains (for cooling purposes) 20 independent sprays, each of which fails with a probability of 0.10. The system fails only if four or more of the sprays fail. What is the probability that the unit will fail?

Solution

Let X denote the number of sprays that fail. The term X has a binomial distribution with $n = 20$ and $p = 0.10$. The probability that the system fails is given by

$$P(X \ge 4) = \sum_{x=4}^{20} \frac{20!}{x!(20-x)!} 0.1^x \, 0.9^{20-x}$$

$$= 1 - \sum_{x=0}^{3} \frac{20!}{x!(20-x)!} 0.1^x \, 0.9^{20-x}$$

$$= 0.1329 = 13.3\%$$

The reader should note the similarity of this example with the previous example.

Illustrative Example 5.6

An engineer's ability to distinguish a natural from a synthetic nano-produced diamond in Africa is tested independently on 10 different occasions. What is the probability of 7 correct identifications if the engineer is only guessing?

Solution
For guessing purposes, the probability of making a correct identification may be reasonably assumed to be 0.5. Let X denote the number of correct identifications. If the engineer is only guessing, X has a binomial distribution with $n = 10$ and $p = 0.5$. The probability of exactly seven correct identifications is

$$P(X = 7) = \frac{10!}{(7!)(3!)}(0.5)^7 (0.5)^3$$
$$= 0.1172 = 11.7\%$$

Illustrative Example 5.7

A procuring agent for an environmental engineering firm is asked to sample a lot of 100 pumps. The sampling procedure calls for the inspection of 20 pumps. If there are any bad pumps, the lot is rejected; otherwise it is accepted. The chief engineer has asked the agent the following question: When there are four bad pumps in the lot, how often would one expect to accept the lot?

Solution
In this problem, $n = 20$ and (since there are four bad pumps in the sample lot) $p = 4/100 = 0.04$. Therefore,

$$P(X = 0) = \frac{20!}{(0!)(20!)}(0.04)^0 (0.96)^{20}$$
$$= 0.442 = 44.2\%$$

Illustrative Example 5.8

A redundant system consisting of 3 operating pumps can survive 2 pump failures in a nuclear power plant. Assume that the pumps are independent with respect to failure and each has a probability of failure of 0.10. What is the reliability of the system?

Solution
The system consists of 3 pumps. Therefore, $n = 3$. The system can survive the failure of 2 pumps. Equation 5.8 should be employed with $r = 2$. The probability of a pump failure is 0.10. Therefore, $p = 0.10$. Substitute these values of n and p in the binomial pdf:

$$f(x) = \frac{3!}{x!(3-x)!}(0.10)^x (0.90)^{3-x}, \quad x = 0,1,2,3$$

Sum the binomial pdf from 0 to r to find the reliability, R:

$$R = \sum_{x=0}^{2} \frac{3!}{x!(3-x)!} 0.10^x 0.90^{3-x}$$
$$= 0.999 = 99.9\%$$

Illustrative Example 5.9

The probability that a rocket hits a target is 0.2. If the rocket is fired 100 times, calculate the expected number (μ) of times the rocket will hit the target. Also calculate the standard deviation σ.

Solution
Assume this probability has a binomial distribution. For this case, $p=0.2$ and $q=0.8$. Therefore

$$\mu = np = (100)(0.2)$$
$$= 20$$

$$\sigma = \sqrt{npq} = \sqrt{(100)(0.2)(0.8)} = \sqrt{16}$$
$$= 4$$

REFERENCES

1. L. Theodore and F. Taylor, *Probability and Statistics*, Theodore Tutorials (originally published by USEPA, RTP, NC), East Williston, NY, 1996.
2. S. Shaefer and L. Theodore, *Probability and Statistics Applications in Environmental Science*, CRC Press/Taylor & Francis Group, Boca Raton, FL, 2007.

6 Multinomial Distribution

INTRODUCTION

Any discussion on the multinomial distribution first requires an introduction to the general probability subject of *permutations* and *combinations*. It is for this reason that this chapter contains two sections: permutations and combinations plus the multinomial theorem.

PERMUTATIONS AND COMBINATIONS

The problem of calculating probabilities of *objects* or *events* in a finite group—defined as the *sample space*—in which equal probabilities are assigned to the elements in the sample space requires counting the elements that make up the events. The counting of such events is often greatly simplified by employing the rules for permutations and combinations.[1,2]

Permutations and combinations deal with the grouping and arrangement of objects or events. By definition, each different ordering in a given manner or arrangement *with* regard to the order of all or part of the objects is called a *permutation*. Alternately, each of the sets that can be made by using all or part of a given collection of objects *without* regard to the order of the objects in the set is called a *combination*. Although permutations or combinations can be obtained with replacement or without replacement, most analyses of permutations and combinations are based on sampling that is performed without replacement, that is, each object or element can be used only once. For each of the two *with/without* pairs (with/without regard to order and with/without replacement), four subsets of two may be drawn. These four are provided in Table 6.1.

Each of the four paired subsets in Table 6.1 is considered in the following text with accompanying examples based on the letters A, B, and C. To personalize this, the reader could consider the options (games of chance) the author faces while on a 1-day visit to a casino. The only three options normally considered are dice (often referred to as craps), blackjack (occasionally referred to as 21), and pari-mutuel (horses, trotters, dogs, and jai alai) simulcasting betting. All three of these may be played during a visit, although playing two or only one is also an option. In addition, the order may vary, and the option may be repeated. Some possibilities include the following:

- Dice, blackjack, and then simulcast wagering
- Blackjack, wagering, and dice
- Wagering, dice, and wagering
- Wagering and dice (the author's usual sequence)
- Blackjack, blackjack (following a break), and dice

TABLE 6.1

Subsets of Permutations and Combinations

Permutations (With Regard to Order)	Combinations (Without Regard to Order)
Without replacement	Without replacement
With replacement	With replacement

In order to simplify the examples that follow, dice, blackjack, and wagering are referred to as *objects* represented by the letters A, B, and C, respectively.

1. Consider a scenario that involves three separate objects, A, B, and C. The arrangement of these objects is called a *permutation*. There are six different orders or permutations of these three objects possible, as follows, while noting that $ABC \neq CBA$.

$$ABC \to BAC \to CAB$$
$$ACB \to BCA \to CBA$$

Thus, BCA would represent blackjack, wagering, and dice.

The number of different permutations of n objects is always equal to $n!$, where $n!$ is normally referred to as factorial n. Factorial n or $n!$ is defined as the product of the n objects taken n at a time and denoted as $P(n,n)$ or $P(n/n)$. Thus,

$$P(n,n) = P(n/n) = n! \qquad (6.1)$$

With three objects, the number of permutations is $3! = 3 \times 2 \times 1 = 6$. Note that $0!$ is 1.

The number of different permutations of n objects taken r at a time is given by

$$P(n,r) = P(n/r) = \frac{n!}{(n-r)!} \qquad (6.2)$$

(*Note:* The permutation term P also appears in the literature as nP_r). For the three objects A, B, and C, taken two at a time, ($n = 3$ and $r = 2$). Thus,

$$P(3,2) = \frac{3!}{(3-2)!} = \frac{(3)(2)(1)}{1} = 6$$

These possible different orders, noting once again that $AB \neq BC$, are

$$AB \rightarrow BC \rightarrow AC$$
$$BA \rightarrow CB \rightarrow CA$$

Consider now a scenario involving n objects in which these can be divided into j sets with the objects within each set being alike. If r_1, r_2,..., r_j represent the number of objects within each of the respective sets, with $n = r_1 + r_2 + \cdots + r_j$, then the number of permutations of the n objects is given by

$$P(n; r_1, r_2,...,r_j) = \frac{n!}{r_1! r_2! ... r_j!} \qquad (6.3)$$

This represents the number of permutations of n objects of which r_1 are alike, r_2 are alike, and so on. Consider, for example, 2As, 1B, and 1C; the number of permutations of these 4 objects is

$$P(4;2,1,1) = \frac{4!}{2!1!1!} = 12$$

The 12 permutations for this scenario of 4 objects are as follows:

AABC ABAC ABCA BAAC BACA BCAA
AACB ACAB ACBA CAAB CABA CBAA

2. Consider the arrangement of the same three objects in (1), but obtain the number of permutations (with regard to order) with replacement, PR. There are 27 different permutations possible.

AAA	BBB	CCC
AAB	BBA	CCB
AAC	BBC	CCA
ABA	BAB	CBC
ABB	BAA	CBB
ACA	BCB	CAC
ACC	BCC	CAA
ABC	BAC	CBA
ACB	BCA	CAB

For this scenario,

$$PR(n,n) = PR(n/n) = (n)^n \qquad (6.4)$$

so that

$$PR(3,3) = (3)^3 = 27$$

For n objects taken r at a time,

$$PR(n,r) = PR(n/r) = (n)^r \qquad (6.5)$$

3. The number of different ways in which one can select r objects from a set of n *without* regard to order (i.e., the order does not count) and without replacement is defined as the number of *combinations, C,* of the n objects taken r at a time. The number of combinations of n objects taken r at a time is given by

$$C(n,r) = C(n/r) = \frac{P(n,r)}{r!} = \frac{n!}{(n-r)!} \qquad (6.6)$$

(*Note:* The combination term C also appears in the literature as C_r^n.)
For the ABC example taken two letters at a time, one has

$$C(3,2) = \frac{3!}{2!1!}$$

$$= \frac{(3)(2)(1)}{(2)(1)(1)}$$

$$= 3$$

The number of combinations becomes

AB   AC   BC

Note that the combination BA is not included since AB=BA for combination orders do not count.
4. The arrangement of n objects, taken r at a time without regard to order and with replacement, is denoted by CR and given by[3]

$$CR(n,r) = CR(n/r) = C(n,r) + (n)^{r-1} \qquad (6.7)$$

with

$$CR(n,n) = CR(n/n) = C(n,n) + (n)^{n-1}; \quad r = n \qquad (6.8)$$

For the ABC example taken two letters at a time, one employs Equation 6.7:

$$CR = C(3,2) + (3)^{2-1}$$

$$= 3 + 3$$

$$= 6$$

The number of combinations becomes

$$
\begin{array}{ccc}
AA & BB & CC \\
AB & AC & BC
\end{array}
$$

Table 6.1 may now be rewritten to include the describing equation for each of the earlier four subsets. This is presented in Table 6.2.

For example, the relative advantages or disadvantages of identifying a laboratory sample

1. With three numbers
2. With three letters

can now be discussed. This is an example of sampling with regard to order and with replacement, because the number may be replaced (reused) or replicated. For this case, Equation 6.5 is employed.

1. Three numbers give the following solution:

$$PR(n,\ r) = PR(n/r) = (n)^r; n = 10, r = 3$$

$$PR(10,3) = (10)(10)(10) = 10^3$$

$$= 1000$$

2. With three letters, the solution is as follows:

$$PR(26,3) = (26)(26)(26) = 26^3$$

$$= 17,576$$

Obviously, the latter choice (three letters) provides greater flexibility, because numerous, that is, more, identification possibilities are available.

TABLE 6.2

Describing Equations for Permutations and Combinations

	Without Replacement	With Replacement	Type
With regard to order	$P(n,r) = P(n/r) = \dfrac{n!}{(n-r)!}$ (Equation 6.2)	$PR(n,r) = PR(n/r) = (n)^r$ (Equation 6.5)	Permutation
Without regard to order	$C(n,r) = C(n/r) = \dfrac{n!}{(n-r)!}$ (Equation 6.6)	$CR(n,r) = CR(n/r) = C(n,r) + (n)^{r-1}$ (Equation 6.7)	Combination

One can also determine the number of possible license plates drawn from 26 letters and 10 integers that consists of 7 symbols. An example is the automobile license plate number of the author: CCY9126. (The car was demolished in an accident caused by a young driver that landed the author in ER.) Note that this calculation again involves a situation in which the order of the symbol does matter and the symbol may be repeated. Based on the information provided, there are 26 letters and 10 integers. Therefore, 36 symbols are available to choose from. The number of permutations of r symbols that can be drawn from a pool of n symbols is again given by

$$PR(n,r) = PR(n/r) = (n)^r$$

For this case, $r=7$ and $n=36$. The number is therefore

$$PR(n,r) = PR(n/r) = (36)^7$$

$$= 7.84 \times 10^{10}$$

Obviously, there are a rather large number of permutations.

MULTINOMIAL THEOREM

Suppose a random variable cannot only assume two values (as in the binomial distribution) but can also fall into any one of n different classes with respective probabilities $p_1, p_2, \ldots, p_n$. Then, for a total of n observations, the probability that $x_1, x_2, \ldots, x_n$ observations will fall into classes $1, 2, \ldots, n$, respectively, is given by the frequency function defined as the *multinomial distribution*:

$$f(x_1, x_2, \ldots, x_n) = \frac{n!}{x_1! x_2! \ldots x_n!} (p_1)^{x_1} (p_2)^{x_2} \ldots (p_n)^{x_n} \tag{6.9}$$

With respect to any variable x_i, the multinomial distribution has a mean and variance of

$$\mu = np \tag{6.10}$$

and

$$\sigma^2 = np(1-p) \tag{6.11}$$

When applied to any variable i, that is, x_i,

$$\mu_i = np_i \tag{6.12}$$

$$\sigma_i^2 = np_i(1-p_i) \tag{6.13}$$

Also note that (once again)

$$n_1 + n_2 + \cdots + n_n = n \tag{6.14}$$

and

$$p_1 + p_2 + \cdots + p_n = 1 \tag{6.15}$$

In addition, if $n=2$, one obtains the binomial distribution discussed in the previous chapter.

Consider the following example. If one die is thrown 12 times, one can calculate the probability of having the numbers 1, 2, 3, 4, 5, and 6 appear twice atop the die after the 12 throws using the multinomial distribution equation. Applying Equation 6.9 leads to

$$P(x_1 = n_1, x_2 = n_2, x_3 = n_3, x_4 = n_4, x_5 = n_5, x_6 = n_6)$$

$$= \frac{n!}{n_1! n_2! n_3! n_4! n_5! n_6!} (p_1)^{x_1} (p_2)^{x_2} (p_3)^{x_3} (p_4)^{x_4} (p_5)^{x_5} (p_6)^{x_6}$$

Substituting

$$P(x_1 = 2, x_2 = 2, x_3 = 2, x_4 = 2, x_5 = 2, x_6 = 2)$$

$$= \frac{12!}{(2!)(2!)(2!)(2!)(2!)(2!)} \left(\frac{1}{6}\right)^2 \left(\frac{1}{6}\right)^2 \left(\frac{1}{6}\right)^2 \left(\frac{1}{6}\right)^2 \left(\frac{1}{6}\right)^2 \left(\frac{1}{6}\right)^2$$

$$= 0.00344 = 0.344\%$$

As expected, the probability is extremely low. Alternatively, one can also calculate the probability of obtaining the numbers 1 and 2 twice and the numbers 5 and 6 four times is given by

$$P(x_1 = 2, x_2 = 2, x_3 = 0, x_4 = 0, x_5 = 4, x_6 = 4)$$

$$= \frac{12!}{(2!)(2!)(0!)(0!)(4!)(4!)} \left(\frac{1}{6}\right)^2 \left(\frac{1}{6}\right)^2 \left(\frac{1}{6}\right)^0 \left(\frac{1}{6}\right)^0 \left(\frac{1}{6}\right)^4 \left(\frac{1}{6}\right)^4$$

$$\approx 0.0001 = 0.01\%$$

Not surprisingly, this number is even lower.

Also consider the following fluid flow example. There are five fans, four pumps, and three small compressors in a shop of a chemical plant. If one of the prime movers is selected from the shop and then returned for later use, calculate the probability that out of 6 prime movers selected, 3 are fans, 2 are pumps, and 1 is a compressor. The probabilities for this application are as follows:

$$P(\text{fan}) = 5/12$$

$$P(\text{pump}) = 4/12$$

$$P(\text{compressor}) = 3/12$$

Applying Equation 6.9 with subscripts 1, 2, and 3 referring to fans, pumps, and compressors, respectively, leads to

$$P(x_1 = 3, x_2 = 2, x_3 = 1) = \frac{6!}{(3!)(2!)(1!)} \left(\frac{5}{12}\right)^3 \left(\frac{4}{12}\right)^2 \left(\frac{3}{12}\right)^1$$

$$= (60)(0.0723)(0.111)(0.25)$$

$$= 0.122 = 12.2\%$$

This fluid flow application could also have been solved by noting that the probability of choosing 3 fans is $(5/12)^3$. Similarly, the probability of choosing 2 pumps is $(4/12)^2$, and of choosing 1 compressor, $(3/12)^1$. Therefore, the probability of choosing 3 fans, 2 pumps, and 1 compressor is *exactly* that order in

$$\left(\frac{5}{12}\right)^3 \left(\frac{4}{12}\right)^2 \left(\frac{3}{12}\right)^1$$

But the same selection can be achieved in various other orders, and the number of these different selection processes is

$$\frac{6!}{(3!)(2!)(1!)}$$

as discussed earlier in the chapter. Thus, the required probability remains

$$\frac{6!}{(3!)(2!)(1!)} \left(\frac{5}{12}\right)^3 \left(\frac{4}{12}\right)^2 \left(\frac{3}{12}\right)^1$$

ILLUSTRATIVE EXAMPLES

Illustrative Example 6.1

How many 4-digit laboratory labels can be formed from the numbers 0 to 9 if a number cannot be repeated in the 4-digit label?

Solution
These are permutations without replacement with $n=10$ and $r=4$. Therefore, Equation 6.2 applies

$$P(n,r) = P(n/r) = \frac{n!}{(n-r)!} \tag{6.2}$$

Substituting

$$P(10,4) = \frac{10!}{6!} = (10)(9)(8)(7)$$
$$= 5,040$$

Note once again that

$$n! = (n)(n-1)...(2)(1)$$

and

$$0! = 1$$

Illustrative Example 6.2

Determine the number of 4-element chemical compounds that can theoretically be generated from a pool of 112 elements. Assume that each element counts only once in the chemical formula and that the order of the elements in the compound matters. An example of a 3-element compound is H_2SO_4 (sulfuric acid) or CH_3OH (methanol). An example of a 4-element compound is $NaHCO_3$.

Solution
As discussed earlier, each different ordering or arrangement of all or part of a number of symbols (or objects) in which the order matters is defined as a *permutation*. There are 112 elements to choose from in this application. For this problem, the describing equation once again is given by the following equation:

$$P(n,r) = P(n/r) = \frac{n!}{(n-r)!} \tag{6.2}$$

Based on the problem statement, $n = 112$ and $r = 4$. Therefore,

$$P(112,4) = \frac{112!}{(112-4)!}$$

$$= (112)(111)(110)(109)$$

$$= 1.49 \times 10^8$$

This is a large number. This number would be further increased if the number of a particular element appearing in the chemical formula was greater than 1. For example, HCN (hydrogen cyanide) vs. C_3H_3N (acrylonitrile). However, a more realistic scenario would involve a calculation in which the order does not matter.

Two points need to be made, one concerning the calculation and the other concerning the chemistry of the compound:

1. For calculations involving large factorials, it is often convenient to use an approximation known as Stirling's formula:

$$n! = (2\pi)^{0.5} e^{-n} n^{n+0.5} \tag{6.2.1}$$

2. For a real-world *viable* compound, the elements involved in this application must be capable of bonding.

Illustrative Example 6.3

Refer to the previous example. Recalculate the number of chemical compounds that can be formed in which the arrangement of symbols is such that the order does not matter, as with a chemical compound. For example, HCN is the same as CNH.

Solution
For this case, the combination equation (Equation 6.6) applies:

$$C(n,r) = C(n/r) = \frac{P(n,r)}{r!} = \frac{n!}{(n-r)!} \tag{6.6}$$

Substituting once again gives

$$C(112,4) = \frac{112!}{(4!)(112-4)!} = \frac{112!}{(4!)(108!)}$$

$$= \frac{(112)(111)(110)(109)}{(4)(3)(2)(1)}$$

$$= 6.21 \times 10^6$$

As expected, the result is lower than that generated in the previous example.

Illustrative Example 6.4

The packing (for a mass transfer unit, e.g., an absorber) produced by a company is 50% rings (A), 30% Tellerettes (B), and 20% saddles (C). In a sample of 5 packings, find the probability P of 2, 1, and 2 of A, B, and C, respectively, appearing in the sample.

Solution
Apply the multinomial distribution equation (Equation 6.9) and substitute

$$P = P(x_1 = 2, x_2 = 1, x_3 = 2) = \frac{5!}{2!1!2!}(0.5)^2(0.3)^1(0.2)^2$$

$$= 0.09 = 9.0\%$$

Illustrative Example 6.5

A manufacturing plant produces, on average, 5 large-sized, 3 medium-sized, and 2 small-sized ball bearings. A sample of 6 ball bearings is drawn. Find the probability that 2 of each size will be drawn.

Solution
For this example,

$$p_1 = 0.5$$

$$p_2 = 0.3$$

$$p_3 = 0.2$$

$$n = 6$$

Apply the multinomial equation in Equation 6.9 and substitute

$$P = P(x_1 = 2, x_2 = 2, x_3 = 2) = \frac{6!}{2!2!2!}(0.5)^2(0.3)^2(0.2)^2$$

$$= 0.0811 = 8.1\%$$

Illustrative Example 6.6

A small biotech lab possesses a large number of hypodermic needles of which 50% are classified as "large diameter" (1), 30% "medium diameter" (2), and 20% "small diameter" (3). If 5 are drawn from the needle lot, what is the probability that 3 will be of a large diameter, none will be medium diameter, and 2 will be small diameter?

Solution

For this example,

$$p_1 = 0.5; \quad n_1 = 3$$
$$p_2 = 0.3; \quad n_2 = 0$$
$$p_3 = 0.2; \quad n_3 = 2$$

Substituting into Equation 6.9 yields

$$P = P(x_1 = 3, x_2 = 0, x_3 = 2) = \frac{5!}{3!0!2!}(0.5)^3 (0.3)^0 (0.2)^2$$

$$= 0.05 = 5.0\%$$

Note that the requirement of a "large number of needles" can be omitted if they are drawn one at a time and then replaced.

Illustrative Example 6.7

A supply shed at a plant contains 5 shell and tube heat exchangers (S&T), 2 double pipe exchangers (DP), and 3 finned exchangers (F).[4] A young engineer examined 6 exchangers (with replacement). Find the probability that she might draw 3 S&T, 1 DP, and 2 F exchangers.

Solution

First note that

$$P(S\&T) = 5/10 = 0.5$$
$$P(DP) = 2/10 = 0.2$$
$$P(F) = 3/10 = 0.3$$

Apply Equation 6.9:

$$P(5,2,3) = \frac{6!}{(3!)(1!)(2!)}(0.5)^3 (0.3)^1 (0.2)^2$$

$$= (60)(0.125)(0.2)(0.09)$$

$$= 0.135 = 13.5\%$$

Illustrative Example 6.8

An industrial plant is currently producing 1% scrap and 5% rework. In effect, 1% cannot be reused at all, while 5% of the product is usable and the remaining 94% is *good* material. A sample of 50 pieces is chosen from a round-the-clock production line. Would you consider 2% scrap and 8% reworkable material as unusual? Clearly state any assumption(s).

Solution

This involves a tricky application of the multinomial distribution, provided in Equation 6.9, that is,

$$P(x_1, x_2, x_3) = \frac{n!}{x_1! x_2! x_3!} (p_1)^{x_1} (p_2)^{x_2} (p_3)^{x_3} \tag{6.9}$$

For this case,

$$p_1 = 0.01$$

$$p_2 = 0.05$$

$$p_3 = 0.94$$

with $n = 50$ and

$$x_1 = (0.02)(50) = 1$$
$$x_2 = (0.08)(50) = 4$$

while

$$x_3 = n - x_1 - x_2 = 50 - 1 - 4$$
$$= 45$$

Assume the sample is considered unusual if $x_1 \geq 1$ and $x_2 \geq 4$. This probability is given by

$$P(x_1 \geq 1, x_2 \geq 4) = \sum_{x_1 \geq 1, x_2 \geq 4} \frac{50!}{x_1! x_2! (50 - x_1 - x_2)!} (0.01)^{x_1} (0.05)^{x_2} (0.94)^{50 - x_1 - x_2}$$

To simplify the calculation, note that

$$P = 1 - \sum P(x_1 = 0, x_2 = 4) + \sum P(x_1 = 1, x_2 = 4)$$

Nine terms need to be evaluated

$$x_1 = 0; \quad x_2 = 0$$
$$x_1 = 0; \quad x_2 = 1$$
$$x_1 = 0; \quad x_2 = 2$$
$$x_1 = 0; \quad x_2 = 3$$
$$x_1 = 0; \quad x_2 = 4$$

and

$$x_1 = 1; \quad x_2 = 0$$
$$x_1 = 1; \quad x_2 = 1$$
$$x_1 = 1; \quad x_2 = 2$$
$$x_1 = 1; \quad x_2 = 3$$

After substitution,

$$P(x_1 \geq 1, x_2 \geq 4) \approx 0.10 \approx 10\%$$

Is this unreasonable? The reader can make the call.

Illustrative Example 6.9

Refer to the previous example. Calculate the probability of the sample containing exactly 2 scrap pieces.

Solution
For this case,

$$p_1 = 0.01$$
$$p_2 = 0.99 \text{ (not scrap)}$$

One can now employ the binomial distribution/theorem, that is,

$$P(x_1 = 2) = \frac{50!}{(2!)(48!)}(0.01)^2(0.99)^{48}$$

$$= (1\,228)(10^{-4})(0.6171)$$

$$= 0.075 = 7.5\%$$

REFERENCES

1. L. Theodore and F. Taylor, *Probability and Statistics*, Theodore Tutorials (originally published by USEPA, RTP, NC), East Williston, NY, 1996.
2. S. Shaefer and L. Theodore, *Probability and Statistics Applications in Environmental Science*, CRC Press/Taylor & Francis Group, Boca Raton, FL, 2007.
3. L. Theodore, personal notes, East Williston, NY, 1980.
4. L. Theodore, *Heat Transfer Applications for the Practicing Engineer*, John Wiley & Sons, Hoboken, NJ, 2012.

7 Hypergeometric Distribution

INTRODUCTION

The hypergeometric distribution is an applicable situation in which a random sample of r items is drawn without replacement from a set of n items. *Without replacement* means that an item is not returned to the set after it is drawn. Recall that the binomial distribution is frequently applicable in cases where the item is drawn *with replacement*.

Suppose that it is possible to classify each of the n items as a *success* or *failure*. Again, the words success and failure do not have the usual connotation. They are merely labels for two mutually exclusive categories into which n items have been classified. Thus, each element of the population may be dichotomized as belonging to one of two disjointed classes.

Let a be the number of items in the category labeled success. Then $n - a$ be the number of items in the category labeled failure. Let X denote the number of successes in a random sample of r items drawn without replacement from the set of n items. Then the random variable X has a hypergeometric distribution, whose probability distribution function (pdf) is specified as follows:

$$f(x) = \frac{\dfrac{a!}{x!(a-x)!}\dfrac{(n-a)!}{(r-x)!(n-a-r+x)!}}{\dfrac{n!}{r!(n-r)!}}, \quad x = 0,1,\ldots,\min(a,r) \qquad (7.1)$$

The term $f(x)$ is the probability of x successes in a random sample of r items drawn without replacement from a set of n items, where a are classified as successes and $n - a$ as failures. The term $\min(a, r)$ represents the smaller of the two numbers a and r, that is, $\min(a, r) = a$ if $a < r$ and $\min(a, r) = r$ if $r \le a$.

Consider the following example. A sample of 5 transistors is drawn at random without replacement from a lot of 1000 transistors, 50 of which are defective. What is the probability that the sample contains exactly three defectives? The number of items in the set from which the sample is drawn is the number of transistors in the lot. Therefore, $n = 1000$. Associate success with drawing a defective transistor and failure with drawing a nondefective one. Determine a, the number of successes in the set of n items. Because 50 of the transistors in the lot are defective, $a = 50$. Also note that the sample is drawn without replacement and that the size of the sample is $r = 5$.

Substituting the values of n, r, and a in the hypergeometric pdf provided in Equation 7.1 gives

$$f(x) = \frac{\dfrac{50!}{x!(50-x)!} \dfrac{950!}{(5-x)!(945+x)!}}{\dfrac{1000!}{5!995!}}, \quad x = 0,1,\ldots,5$$

Also, substitute the appropriate value of X above to obtain the required probability.

$$P(\text{sample contains exactly three defectives}) = P(X = 3)$$

Therefore,

$$P(X=3) = \frac{\dfrac{50!}{3!(47)!} \dfrac{950!}{2!(945+3)!}}{\dfrac{1000!}{5!995!}}$$

$$= 0.0011 = 0.11\%$$

The hypergeometric distribution can also be arrived at through the use of combinations (see also Chapter 5 on the binomial distribution). Using this approach, the total number of possible selections of r elements of n is $C(n, r)$. Furthermore, x good elements may be chosen out of the total of a good elements in a total of $C(a, x)$ combinations. For each such combination, it is also possible to select $r - x$ of $n - a$ bad elements in $C(a, x)\, C(n - a, r - x)$. Because each selection possibility is equally likely, the probability of picking exactly x good elements is

$$f(x) = \frac{C(a,x)C(n-a,r-x)}{C(n,r)} \tag{7.2}$$

which upon substitution of C (see Chapter 5) once again is the expression given in Equation 7.1 for the probability density function of the hypergeometric distribution. As with the binomial distribution, also note that the Equation 7.2 can be rewritten as follows:

$$f(x) = \frac{\dbinom{a}{x}\dbinom{n-a}{r-x}}{\dbinom{n}{r}} \tag{7.3}$$

Consider the following scenario. A machine ship contains six square bolts and four hexagonal bolts. A bolt is chosen at random and observed, but the bolt is not replaced. Find the probability that after five selections, three square bolts will be chosen. As noted above, this example may be solved by either of two methods.

Employing the hypergeometric approach using Equation 7.1, one has $a=6$, $n=10$, $r=5$, $x=3$. Thus, the probability is

$$P(X=3) = \frac{\binom{6}{3}\binom{4}{2}}{\binom{10}{2}}$$

$$= \frac{10}{21}$$

$$= 0.45 = 45\%$$

Quality control engineers devise acceptance sampling plans for the inspection of incoming lots of merchandise. Sampling plans are evaluated on the basis of protection afforded against (a) accepting a bad lot and (b) rejecting a good lot. The probability of accepting a bad lot is called *consumer risk* and the probability of rejecting a good lot is called *producer risk*. This plan is reviewed in the following two paragraphs.

Acceptance sampling is thus a set of methods for accepting or rejecting a lot of N items on the basis of inspecting a sample of size n items selected at random from the lot. In practice, acceptance sampling requires that three numbers be specified.

1. $N=$ the number of items from which the sample is to be taken
2. $n=$ the number of items to be sampled, and
3. $c=$ the maximum allowable number of defective items in a sample of size n; it is often referred to as the *acceptance number*

In a sampling plan, the lot is rejected if the sample contains more than c defective items.

Consider the following sampling plan. A quality control engineer inspects a random sample of 3 calculators drawn without replacement from each incoming lot of 25 calculators. A lot is accepted only if the sample contains no defectives. If any defectives are found, the entire lot is rejected and any cost is charged to the vendor. What is the probability of

1. Accepting a lot containing 10 defective calculators?
2. Inspecting an entire lot containing only one defective calculator?

A recurrence relation for the hypergeometric probabilities is available. Note that a recurrence relation for the binomial coefficients was presented in Chapter 5. Thus, it is not surprising that there is also a recurrence relation for the hypergeometric probabilities. The recurrence relation is given by

$$f(x+1) = \left[\frac{(n-x)(a-x)}{(x+1)(N-a-n+x+1)}\right]f(x) \qquad (7.4)$$

where

$$\max\left(0, n-(N-a)\right) \le x \le \min(n,a) \tag{7.5}$$

This equation may also be written as

$$f(x+1) = \left[RH(x)\right]f(x) \tag{7.6}$$

where the term within the brackets is the recurrence coefficient

$$RH(x) = \left[\frac{(n-x)(a-x)}{(x+1)(N-a-n+x+1)}\right] \tag{7.7}$$

For example, one can calculate the probabilities for the number of spades in a five-card poker hand using the recurrence relation presented in Equation 7.4. For this case,

$$n=5$$

$$N=52$$

$$a=13$$

$$x=0, 1, 2, 3, 4, \text{ and } 5$$

Therefore, employing the "combination" form of the hypergeometric distribution,

$$f(0) = \frac{\binom{13}{0}\binom{39}{5}}{\binom{52}{5}} = 0.2215$$

plus

$$RH(x) = \left[\frac{(5-x)(13-x)}{(x+1)(35+x)}\right], \qquad x=0,1,\ldots$$

Therefore, for $x = 0$

$$f(1) = \left[\frac{(5)(13)}{35}\right]0.2215 = 0.4114$$

$$f(2) = \left[\frac{(4)(12)}{(2)(36)}\right]0.4114 = 0.2743$$

$$f(3) = \left[\frac{(3)(11)}{(3)(37)} \right] 0.2743 = 0.0815$$

$$f(4) = \left[\frac{(2)(10)}{(4)(38)} \right] 0.0815 = 0.0107$$

and (for a flush)

$$f(5) = \left[\frac{(1)(9)}{(5)(39)} \right] 0.0107 = 0.0005$$

Are there differences between the hypergeometric and binomial distributions? The hypergeometric distribution is obviously a special case of the binomial distribution when applied to finite populations. In particular, the hypergeometric distribution approaches the binomial distribution when the population size approaches infinity. Note, however, that others have claimed that the binomial distribution is a special case of a hypergeometric distribution.

Finally, the hypergeometric distribution is applicable in situations similar to those when the binomial distribution is used, except that samples are taken from a *small* population. Examples arise in sampling from small numbers of chemical, medical, and environmental samples, as well as from manufacturing lots.

ILLUSTRATIVE EXAMPLES

Illustrative Example 7.1

A pillbox contains 24 drug tablets, 3 of which are contaminated. If a sample of 6 is chosen at random from the pillbox, what is the probability that 0 will be contaminated?

Solution
First note that this involves sampling *without* replacement. Employing the combination equation provided in the introductory section of this chapter, Equation 7.2 yields

$$f(x) = \frac{C(a,x)C(n-a,r-x)}{C(n,r)}, \quad a = 3, r = 6, n = 24$$

Therefore,

$$f(x) = \frac{C(3,x)C(21,6-x)}{C(24,6)}, \quad 0 \le x \le 3$$

There are only 4 values X can assume: 0, 1, 2, and 3. Substituting $X=0$ gives

$$P(X = 0) = \frac{C(3,0)C(21,6)}{C(24,6)}$$

$$= \frac{\left(\dfrac{3!}{0!3!}\right)\left(\dfrac{21!}{6!15!}\right)}{\left(\dfrac{24!}{6!18!}\right)}$$

$$= \frac{(1)(54,264)}{(134,596)}$$

$$= 0.4032 = 40.3\%$$

Illustrative Example 7.2

Refer to the previous example. Resolve the problem if the number of contaminated pills in the sample of 6 is

1. One
2. Two
3. Three

Solution

1. For 1 pill,

$$P(X = 1) = \frac{C(3,1)C(21,5)}{C(24,6)}$$

$$= 0.45356 = 45.36\%$$

2. For 2 pills,

$$P(X = 2) = \frac{C(3,2)C(21,4)}{C(24,6)}$$

$$= 0.13340 = 13.34\%$$

3. For 3 pills,

$$P(X = 3) = \frac{C(3,3)C(21,3)}{C(24,6)}$$

$$= 0.00988 = 1\%$$

Note that out of necessity

$$\sum_{x=0}^{3} P(X) = 1$$

Illustrative Example 7.3

Refer to Illustrative Example 7.1. Resolve the problem if the sample size chosen is 8, rather than 6.

Solution

The only difference in this problem is that $r=8$. Therefore, the describing equation becomes

$$P(X=0) = \frac{C(3,0)C(21,8)}{C(24,8)}$$

$$= \frac{\left(\dfrac{3!}{0!3!}\right)\left(\dfrac{21!}{8!13!}\right)}{\left(\dfrac{24!}{8!16!}\right)}$$

$$= \frac{(1)(203,490)}{(735,471)}$$

$$= 0.277 = 27.7\%$$

As expected, the probability decreases.

Illustrative Example 7.4

Refer to the previous example. Resolve the problem if the sample size chosen is 2 rather than 8.

Solution

The describing hypergeometric equation now becomes

$$P(X=0) = \frac{C(3,0)C(21,2)}{C(24,2)}$$

$$= \frac{(1)(210)}{(276)}$$

$$= 0.761 = 76.1\%$$

The binomial equation prediction is

$$P(X=0) = (0.875)^2$$

$$= 0.765 = 76.5\%$$

As expected, the agreement is excellent.

Illustrative Example 7.5

A quality control engineer inspects a random sample of 3 vacuum pumps drawn without replacement from each incoming lot of 25 vacuum pumps. A lot is accepted only if the sample contains no defectives. If any defectives are found, the entire lot is inspected, and the cost charged to the vendor. Develop the hypergeometric distribution equation for this quality test.

Solution
Each lot consists of 25 vacuum pumps. Therefore, $n=25$. Once again, associate success with a defective vacuum pump. Determine, a, the number of successes in the set of n items. For this problem, $a=10$. A sample of three vacuum pumps is drawn without replacement from each lot; therefore, $r=3$.

Substitute the above values of n, r, and a in the pdf of the hypergeometric distribution, that is, Equation 7.1.

$$f(x) = \frac{\dfrac{10!}{x!(10-x)!} \dfrac{15!}{(3-x)!(12+x)!}}{\dfrac{25!}{(3!)(22!)}}, \quad x = 0,1,2,3$$

Illustrative Example 7.6

Refer to the previous example. What is the probability of accepting an entire lot containing only 1 defective vacuum pump?

Solution
The pdf with only 1 defective pump is

$$f(x) = \frac{\dfrac{1!}{x!(1-x)!} \dfrac{24!}{(3-x)!(21+x)!}}{\dfrac{25!}{(3!)(22!)}}, \quad x = 0,1$$

For no defective vacuum pumps,

$$n = 25, \quad a = 1, \quad r = 3, \quad x = 0$$

Therefore,

$$P(X = 0) = \frac{\dfrac{1!}{(0!)(1!)} \dfrac{24!}{(3!)(21!)}}{\dfrac{25!}{(3!)(22!)}}, \quad x = 0,1$$

$$= \frac{\dfrac{24!}{(3!)(21!)}}{\dfrac{25!}{(3!)(22!)}}$$

$$= \frac{2024}{2300}$$

$$= 0.88 = 88\%$$

Illustrative Example 7.7

An order to manufacture a special type of water pollution monitor was recently received. The total order consists of 25 monitors. Five of these monitors are to be selected at random and initially life tested. The contract specifies that if not more than 1 of the 5 monitors fails a specified life test, the remaining 20 will be accepted. Otherwise, the complete lot will be rejected. What is the probability of lot acceptance if exactly 4 of the 25 submitted monitors are defective?

Solution

The lot will be accepted if the sample of 5 contains either 0 or 1 of the 4 defective monitors. The probability of this happening is given by the hypergeometric distribution as

$$P(X \le 1) = \frac{C(4,x)C(21,5-x)}{C(24,5)}$$

$$= P(X = 0) + P(X = 1)$$

$$= \frac{\left(\frac{4!}{(4!)(0!)}\right)\left(\frac{21!}{(5!)(16!)}\right)}{\left(\frac{25!}{(20!)(5!)}\right)} + \frac{\left(\frac{4!}{(3!)(1!)}\right)\left(\frac{21!}{(4!)(17!)}\right)}{\left(\frac{25!}{(20!)(5!)}\right)}$$

$$= \frac{(1)(20,349)}{(53,130)} + \frac{(4)(5,985)}{(53,130)}$$

$$= \frac{20,349}{53,130} + \frac{23,940}{53,130}$$

$$= 0.383 + 0.451$$

$$= 0.834 = 83.4\%$$

Illustrative Example 7.8

Refer to the previous example. Outline how the calculation would be affected if 6 monitors are selected for life testing.

Solution

The hypergeometric distribution again applies with the r term now equal to 6 (not 5). Therefore,

$$P(X \le 1) = \sum_{x=0}^{1} \frac{C(4,x)C(21,6-x)}{C(25,6)}$$

Solving this problem is left as an exercise for the reader.

Illustrative Example 7.9[3]

A lot consists of 100 heat exchanger tubes. The sampling calls for the inspection of 20 tubes. If there are any bad tubes, the lot is rejected; otherwise, it is accepted. Answer the following three questions.

1. When there are 5 bad tubes in the lot, how often would one expect to accept it?
2. When there are 10 bad tubes in the lot, how often would one reject it?
3. Suppose 1 bad tube was allowed in the sample. What kind of protection would the purchaser have?

Solution

For (1),

$$P(X=0) = \frac{\binom{5}{0}\binom{95}{20}}{\binom{100}{20}} = 0.3193$$

Thus, one would accept the lot about 32% of the time.

For (2),

$$n = 10$$

so that

$$P(X=0) = \frac{\binom{10}{0}\binom{90}{20}}{\binom{100}{20}} = 0.0951$$

But, if one found 0 defective ($X=0$), the tubes would be accepted. The probability of rejection is therefore $1 - 0.0951 = 0.9049 \approx 90\%$.

For (3), one desires the probability of accepting the tubes with 5 defective if 0 or 1 defective in the sample was allowed. Thus, $n=5$, and

$$P(X \leq 1) = P(X=0) + P(X=1)$$

$$= \frac{\binom{5}{0}\binom{95}{20}}{\binom{100}{20}} + \frac{\binom{5}{1}\binom{95}{19}}{\binom{100}{20}}$$

$$= 0.3193 + 0.4201$$

$$= 0.7394 = 74\%$$

This essentially represents the consumer's risk.

REFERENCES

1. L. Theodore and F. Taylor, *Probability and Statistics*, Theodore Tutorials (originally published by USEPA, RTP, NC), East Williston, NY, 1996.
2. S. Shaefer and L. Theodore, *Probability and Statistics Applications in Environmental Science*, CRC Press/Taylor & Francis Group, Boca Raton, FL, 2007.
3. L. Theodore, Class notes, source unknown, East Williston, NY, 1967.

8 Poisson Distribution

INTRODUCTION

The probability distribution function (pdf) of the Poisson (named after Simeon Poisson) distribution can be derived by taking the limit of the binomial pdf as $n \to \infty$, $P \to 0$, and $nP = \mu$ remains constant. The Poisson pdf is given by

$$f(x) = \frac{e^{-\mu}\mu^x}{x!}, \quad x = 0,1,2,\ldots \tag{8.1}$$

The term $f(x)$ is the probability of x occurrences of an event that occurs on the average μ times per unit of space or time. Both the mean and the variance of a random variable X having a Poisson distribution are μ. Interestingly, the Poisson parameter is both the mean and variance of the distribution.[1,2]

Examining Equation 8.1, if any "incident" occurred at an *average* rate of λ incidents per time, then the probabilities of 0, 1, 2,... incidents occurring in any one unit of time would be expected to equal the successive terms in Equation 8.1 of the Poisson distribution, that is,

$$e^{-\lambda}\left[1 + \lambda + \frac{\lambda^2}{2!} + \frac{\lambda^3}{3!}\right] \tag{8.2}$$

Three bar charts of the Poisson distribution for various values of λ equal to 1, 2, and 5 are presented in Figure 8.1.

In the 155 years that the Kentucky Derby has been run, only 68 horses that entered the race have been undefeated. What is the probability that two undefeated horses will be entered in the next "Run for the Roses?" Assume the Poisson distribution may be applied. For this exercise,

$$\mu = \frac{68}{155}$$

$$= 0.439$$

$$n = 2$$

$$P(X = 2) = P(2) = \frac{e^{-0.439}(0.439)^2}{2!}$$

$$= 0.062 = 6.2\%$$

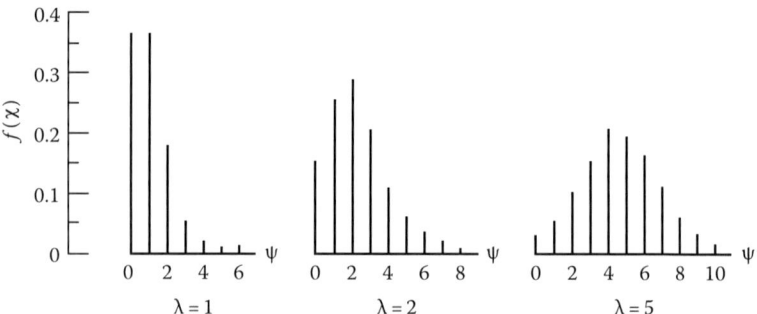

FIGURE 8.1 Poisson distribution for three values of λ.

In effect, the race will be run with two undefeated horses, approximately six times every 100 years. (As some of the readers know, the author is an avid supporter of the "Sport of Kings.")

Consider the following example. If λ is the failure rate (per unit of time) of each component of a system, then λt is the average number of failures for a given unit of time. The probability of x failures in the specified unit of time is obtained by substituting $\mu = \lambda t$ in Equation 8.1 to obtain

$$f(x) = \frac{e^{-\lambda t}(\lambda t)^{x}}{x!}, \quad x = 0, 1, 2, \ldots \tag{8.3}$$

Suppose, for example, that in a certain country the average number of airplane crashes per year is 2.5. What is the probability of 4 or more crashes occurring during the next year? Substituting $\lambda = 2.5$ and $t = 1$ in Equation 8.3 yields

$$f(x) = \frac{e^{-2.5}(2.5)^{x}}{x!}, \quad x = 0, 1, 2, \ldots$$

as the pdf of X, the number of airplane crashes in 1 year. The probability of four or more airplane crashes next year is then

$$P(X \geq 4) = 1 - \sum_{x=0}^{3} \frac{e^{-2.5}(2.5)^{x}}{x!}$$

$$= 1 - (0.0821 + 0.205 + 0.257 + 0.214)$$

$$= 1 - 0.76$$

$$= 0.24 = 24\%$$

This is obviously not an acceptable risk.

In an even simpler example, assume 2% of the thermometers made by a factory for a waste incinerator are defective. Find the probability P that there are 3 defective thermometers in a sample of 100 items. One first notes that the binomial distribution with $n = 100$ and $p = 0.02$ applies. However, since p is small, one can use a Poisson approximation with $\lambda = np = 2$. Thus,

$$P(X = 3) = \frac{e^{-2}2^3}{3!}$$

$$= \frac{(8)(0.135)}{6}$$

$$= 0.180 = 18\%$$

In another example, suppose that the average number of breakdowns of personal computers during 1000 h of operation of a computer center is 3. What is the probability of no breakdowns during a 10 h work period? Note that the given average is the average number of breakdowns during 1000 h. The unit of time associated with the given average is 1000 h. The probability required is the probability of no breakdowns during a 10 h period, and the unit of time connected with the required probability is 10 h. If 10 is divided by 1000, the result is 0.01 so that in a 10 h period, there are 0.01 time periods of 1000 h duration. The given average of breakdown is 3. Multiplication of 3 by 0.01 yields 0.03, the average number of occurrences during a 10 h time period, denoted as μ. This value of μ may be substituted in the Poisson pdf:

$$f(x) = \frac{e^{-0.03}0.03^x}{x!}, \quad x = 0,1,2,\ldots$$

One may now substitute for x, the number of occurrences whose probability is required. The probability of no breakdowns in a 10 h period is

$$P(X = 0) = \frac{e^{-0.03}0.03^0}{0!}$$

$$= 0.97 = 97\%$$

The Poisson pdf can be used to approximate probabilities obtained from the binomial pdf given earlier when n is large and p is small. In general, good approximations will result when n exceeds 10 and p is less than 0.05. When n exceeds 100 and np is less than 10, the approximation will generally be excellent. Table 8.1 compares binomial and Poisson probabilities for the case of $n = 20$ ($p = 0.05$) and $n = 100$ ($p = 0.01$).

The Poisson distribution plays a role in operations research, particularly in the analysis of *queueing systems* where a queueing system is defined as any service

TABLE 8.1

Binomial/Poisson Distribution Comparison

X	Binomial ($n=20, p=0.05$)	Poisson ($n=20, p=0.05$)	Binomial ($n=100, p=0.01$)	Poisson ($n=100, p=0.01$)
0	0.3585	0.3679	0.3660	0.3679
1	0.3774	0.3679	0.3697	0.3679
2	0.1887	0.1839	0.1849	0.1839
3	0.0596	0.0613	0.0610	0.0613
4	0.0133	0.0153	0.0149	0.0153
5	0.0022	0.0031	0.0029	0.0031
6	0.0003	0.0005	0.0005	0.0005
≥7	0.0000	0.0001	0.0001	0.0001

facility to which *customers* or *jobs* arrive, receive service, and then depart.[3] Note that the word customer (in a general sense) can refer to

1. A telephone call arriving at a telephone exchange
2. An order for a component stocked in a warehouse
3. A broken component brought to a repair shop, or
4. A packet of digital data arriving at some node in a complex computer network

The *service time S* is defined as the amount of time required to service the customer. The length of a telephone call and the time to repair a broken component are examples of service times. In most typical applications, both the customer arrivals and their service times are random variables. Queueing systems are classified according to three factors:

1. The *input process*, which denotes the probability distribution of the customer arrivals
2. The *service distribution*, which denotes the probability distribution of the service time, and
3. The *queueing discipline*, which refers to the order of service

Rosenkrantz[3] provides additional information and illustrative examples.

A recurrence relation for the Poisson distribution has some useful applications. The recurrence coefficient *RP* is given by

$$RP(x) = \frac{p(x+1; \lambda)}{p(x; \lambda)} = \frac{\lambda}{x+1}, \quad x = 0, 1, \dots \tag{8.4}$$

which may be rearranged to give

$$(x+1)p(x+1; \lambda) = \lambda p(x; \lambda) \, (x = 0, 1, \dots) \tag{8.5}$$

ILLUSTRATIVE EXAMPLES

Illustrative Example 8.1

The New York Racing Organization (NYRO) has reported that the average number of times a racehorse breaks down and has to be euthanized during the running of 10,000 races is 30. What is the probability that no horses breakdown over a 10-race period?

Solution

As with the example in the Introduction section of this chapter, the unit of "time" associated with the probability to be calculated is 10 races. If 10 is divided by 10,000, the result is 0.001, that is, the 10-race period represent 0.001 races of a total of 10,000 races. Because the average number of horses breaking down in the 10,000-race period is 30, multiplication of 30 by 0.001 yields 0.03. This represents μ, the average number of breakdowns during a 10-race period. Substituting this into Equation 8.1 yields

$$f(x) = \frac{e^{-0.03}0.03^x}{x!}, \quad x = 0, 1, 2, \ldots$$

as the pdf of X, the number of breakdowns in a 10-race period. Then, the probability of no breakdowns in a 10-race period is

$$P(X = 0) = e^{-0.03} = 0.97$$

The reader should determine why this result is identical to that calculated in the Introduction section of this chapter.

Illustrative Example 8.2

Microscopic slides of a certain culture of microorganisms contain on an average 20 microorganisms per square centimeter. After treatment by a chemical, 1 cm² is found to contain only 10 such microorganisms. If the treatment had no effect, what would be the probability of finding 10 or fewer microorganisms in a given square centimeter?

Solution

Let X denote the number of microorganisms in 1 cm². If the chemical treatment had no effect, X has a Poisson pdf with $\mu = 20$. The probability of 10 or fewer microorganisms in a given square centimeter is

$$P(X \leq 10) = \sum_{x=0}^{10} e^{-20}(20)^x$$

$$= P(X = 0) + P(X = 1) + \cdots + P(X = 10)$$

Longhand calculation leads to

$$P(X \leq 10) = 0.0128 = 1.28\%$$

For example,

$$P(X = 0) = 2 \times 10^{-9} \text{ (as expected)}$$

$$P(X = 1) = 4.1 \times 10^{-8}$$

$$P(X = 9) = 0.00291 = 0.3\%$$

$$P(X = 10) = 0.0058 = 0.6\%$$

Illustrative Example 8.3

The average number of defective welds detected at the final examination of the tail section of an aircraft is 5. What is the probability of detecting at least 1 defective weld during the final examination of the tail section?

Solution

The given average is the average number of defective welds detected at the final inspection of an aircraft tail section. The unit of space connected with the given average is the space occupied by the tail section of the aircraft. Therefore, the associated unit of space is the same as the space occupied by the tail section of the aircraft. Because both the unit of space connected with the given average and the unit of space connected with the required area are the same, their quotient is 1. The probability required is the probability of detecting at least 1 defective weld at the final inspection of the tail section. The given average number of defectives is 5. Multiplication by 1 yields the value of μ as 5. Substitute this value of μ in the Poisson pdf.

$$f(x) = \frac{e^{-5}(5)^x}{x!}, \quad x = 0, 1, 2, \ldots$$

Substitute for x, the number of occurrences whose probability is required. The probability of detecting at least 1 defective weld is

$$P(X \geq 1) = 1 - P(X = 0)$$

$$= 1 - \frac{e^{-5}(5)^0}{0!}$$

$$= 0.9933 = 99.33\%$$

Illustrative Example 8.4

Refer to the previous example. Discuss how an increase in the number of defective welds would be handled.

Solution

The Poisson pdf gives the probability of x occurrences over the unit of space associated with the required probability. As indicated earlier, one must substitute for x in the pdf the number occurrences whose probability is required. However, more than one value of x may have to be substituted if the required probability is the sum of the probabilities for several values of x.

Illustrative Example 8.5

The probability that U.S. citizens aged 72–73 (the age of the author nearly 10 years ago) will die within the year is 0.0417. With a group of 100 such individuals, what is the probability that exactly 5 will die within the year?

Solution

For this problem,

$$p = 0.0417$$
$$n = 100$$
$$\mu = pn = (100)(0.0417)$$
$$= 4.17$$
$$\chi = 5$$
$$P(5) = \frac{(4.17)^5}{5!e^{4.17}}$$
$$= \frac{1260}{(120)(64.7)}$$
$$= 0.162 = 16.2\%$$

Illustrative Example 8.6

Over the last 10 years, a local hospital reported that the number of deaths per year due to temperature inversions (air pollution) was 0.5. What is the probability of exactly 3 deaths in a given year?

Solution

For this problem,

$$P(X = 3) = \frac{e^{-0.5}(0.5)^3}{3!}$$
$$= 0.0126 = 1.26\%$$

Illustrative Example 8.7

Refer to the previous example. Calculate the annual probability of three or more deaths being attributed to temperature inversions.

Solution

For this case,

$$P(X \geq 3) = \sum_{x=3}^{\infty} \frac{e^{-0.5}(0.5)^x}{x!}$$

$$= 1 - \sum_{x=0}^{2} \frac{e^{-0.5}(0.5)^x}{x!}$$

$$= 1 - 0.60065 - 0.3033 - 0.0758$$

$$= 0.0227 = 2.3\%$$

Illustrative Example 8.8

Consider a standby redundancy system with 1 operating unit and 1 on standby, that is, a system that can survive one failure. If the failure rate is 2 units per year, what is the 6-month reliability of the system?[4]

Solution

As discussed earlier, a standby redundant system is one in which one unit is in the operating mode and n identical units are in standby mode. Unlike a parallel system in which all units in the system are active, the standby units are inactive in the standby redundancy system. If all the units have the same failure rate in the operating mode, unit failures are independent of each other, standby units have a 0 failure rate in the standby mode, and there is a perfect switchover to a standby unit when the operating unit fails, then the reliability R of the standby redundancy system is given by

$$R = \sum_{x=0}^{n} \frac{e^{-\mu}\mu^x}{x!} \tag{8.6}$$

This represents the probability of n or fewer failures in a time period on which the average number of failures is μ.

The required reliability is a 6-month reliability. Therefore, the associated time period is 6 months. The given failure rate is 2 units per year. Therefore, the associated time period is 1 year or 12 months. Note that 6 divided by 12 yields 1/2. Multiply this result by the failure rate to obtain μ.

$$\mu = (1/2)(2) = 1$$

Substitute the value of μ in the Poisson pdf:

$$f(x) = \frac{e^{-1}(1)^x}{x!}, \quad x = 0, 1, 2, \ldots$$

Identify n, the number of units in standby mode, $n = 1$.

Compute the required reliability by summing the Poisson pdf from 0 to n. The 6-month reliability is given by

$$R = \sum_{x=0}^{1} \frac{e^{-1}(1)^x}{x!}$$

$$= 0.368 + 0.368$$

$$= 0.736 = 73.6\%$$

Illustrative Example 8.9

Bacteria are known to be present in a source of liquid with a mean number of 3 per cubic centimeter. Ten 1 cm³ test tubes are filled with the liquid. Calculate the probability that all 10 test tubes will show growth (i.e., contain at least one bacterium each).

Solution

Let X denote the number of bacteria in 1 cm³. Substituting $\mu = 3$ in the Poisson pdf gives:

$$f(x) = \frac{e^{-3}(3)^x}{x!}, \quad x = 0, 1, 2, \ldots$$

The probability that a test tube will show growth is

$$P(X \geq 1) = 1 - P(X = 0)$$

$$= 1 - e^{-3}$$

$$= 1 - 0.0498$$

$$= 0.9502 = 95\%$$

The probability that all 10 test tubes will show growth is

$$P(X = 10) = (0.9502)^{10}$$

$$= 0.60 = 60\%$$

Illustrative Example 8.10

Assume the number of particles emitted by a radioactive substance has a Poisson distribution with an average emission of 1 particle per second.

1. Find the probability that at the most 1 particle will be emitted in 3 s.
2. How low an emission rate would be required to make the probability of the emission of at most 1 particle in 3 s at least 0.80?

Solution

1. Let X denote the number of radioactive particles emitted in 3 s. Because $\lambda = 1$ and $t = 3$, $\mu = 3$. Therefore, the pdf of X is

$$f(x) = \frac{e^{-3}(3)^x}{x!}, \quad x = 0, 1, 2, \ldots$$

The probability that at most 1 particle will be emitted in 3 s is

$$P(X \le 1) = \sum_{x=0}^{1} \frac{e^{-3}(3)^x}{x!}$$

$$= P(X = 0) + P(X = 1)$$

$$= e^{-3} + e^{-3}(3)$$

$$= 0.0498 + 0.1494$$

$$= 0.1992 = 19.9\%$$

2. Let λ be the emission rate per second. Then the pdf of X is

$$f(x) = \frac{e^{-3\lambda}(3\lambda)^x}{x!}$$

In addition,

$$P(X \le 1) = \sum_{x=0}^{1} \frac{e^{-3\lambda}(3\lambda)^x}{x!}$$

$$= P(X = 0) + P(X = 1)$$

$$= e^{-3\lambda}(1 + 3\lambda)$$

For the probability of emission of at most 1 particle in 3 s to be at least 0.80, λ, the emission rate per second, must be such that

$$e^{-3\lambda}(1 + 3\lambda) \ge 0.80$$

This will be solved using a trial and error method.

For $\lambda = 0.3$,

$$e^{-3\lambda}(1 + 3\lambda) - 0.80 = -0.0275$$

For $\lambda = 0.2$,

$$e^{-3\lambda}(1 + 3\lambda) - 0.80 = +0.0781$$

Using Newton's method or the equivalent, 0.275 is obtained as an improved approximation to the root of the equation

$$e^{-3\lambda}(1+3\lambda)-0.80 = 0$$

Therefore, an emission rate per second of 0.275 or less is required to make the probability of the emission of at most 1 particle in 3 s at least 0.80.

Illustrative Example 8.11

The number of hazardous waste trucks arriving daily at a certain hazardous waste incineration facility has a Poisson distribution with a parameter $\mu=2.5$. Present facilities can accommodate 3 trucks a day. If more than 3 trucks arrive in a day, the trucks in excess of 3 must be sent elsewhere. On a given day, what is the probability of having to send a truck elsewhere?

Solution

Let X denote the number of trucks on a given day. Then

$$P(\text{sending trucks elsewhere}) = P(X \geq 4)$$

This may also be written as

$$P(X \geq 4) = 1 - P(X < 4) = 1 - P(X \leq 3)$$
$$P(X \geq 4) = 1 - P(X = 0) - P(X = 1) - P(X = 2) - P(X = 3)$$

Employing the Poisson pdf with $\mu=2.5$ gives

$$P(X \geq 4) = 1 - \left[e^{-2.5} + 2.5e^{-2.5} + \frac{(2.5)^2 e^{-2.5}}{2} + \frac{(2.5)^3 e^{-2.5}}{6} \right]$$

$$= 1 - (0.0821 + 0.2052 + 0.2565 + 0.2138)$$

$$= 0.2424 = 24.24\%$$

REFERENCES

1. L. Theodore and F. Taylor, *Probability and Statistics*, Theodore Tutorials (originally published by USEPA, RTP, NC), East Williston, NY, 1996.
2. S. Shaefer and L. Theodore, *Probability and Statistics Applications in Environmental Science*, CRC Press/Taylor & Francis Group, Boca Raton, FL, 2007.
3. W. Rosenkrantz, *Probability and Statistics for Science, Engineering, and Finance*, CRC Press/Taylor & Francis Group, Boca Raton, FL, 2009.
4. L. Theodore, *Chemical Engineering: The Essential Reference*, McGraw-Hill, New York, NY, 2013.

Section III

Continuous Probability Distributions

Believing as I do, in the continuity of nature...

John Tyndall (1820–1893)

This section contains five chapters.

These continuous probability distributions generally are based on time (which is continuous) as the independent variable.

Which chapter has received the greatest attention from a risk analysis/assessment perspective? This is a tough question to answer, but the author would rank them in the following order: Weibull (most attention), normal, exponential, and log-normal (least attention). As noted earlier, other than the log-normal distribution, time is almost always the independent variable.

9 Normal Distribution

INTRODUCTION

The initial presentation in this chapter on normal distributions will focus on failure rate, but can be simply applied to all other applications involving normal distributions. When T, time to failure, has a normal distribution with mean μ and variance σ, its probability distribution function (pdf) is given by

$$f(t) = \frac{1}{\sqrt{2\pi}\sigma} \exp\left[-\frac{1}{2}\left(\frac{t-\mu}{\sigma}\right)^2 \right]; \quad -\infty < t < \infty \tag{9.1}$$

The graph of $f(t)$ is the familiar bell-shaped curve shown in Figure 9.1.[1,2]

There are two points at which the curve in Figure 9.1 changes shape in the normal curve. The curve changes at the first point, from being concave with respect to the horizontal axis, to being convex at the first point. At the second point, it again changes to concavity. These are known as *points of inflection*. Interestingly, the points of inflection are located at an interval σ from the mean, that is, at $\bar{x} \pm \sigma$, or 68% of the area under the curve symmetrically displaced from the mean.

If a variable T is normally distributed with mean μ and standard deviation σ, then the random variable $(T - \mu)/\sigma$ is also normally distributed with a mean of 0 and standard deviation 1. The term $(T - \mu)/\sigma$ is called a *standard normal variable* that is represented by Z, not to be confused with the failure rate $Z(t)$ to be discussed in more detail later.

Table 9.1 is a tabulation of areas under a standard normal curve to the right of z_0 for *nonnegative* values of z_0. Probabilities about a standard normal variable Z can be determined from this table. For example,

$$P(Z > 1.54) = 0.062$$

is obtained directly from Table 9.1 as the area to the right of 1.54. As presented in Figure 9.2, the symmetry of the standard normal curve about zero implies that the area to the right of zero is 0.5, and the area to the left of zero is 0.5.

Plots demonstrating the effect of μ and σ on the bell-shaped curve are provided in Figures 9.3 and 9.4. Consequently, one can deduce from Table 9.1 and Figure 9.2 that

$$P(0 < Z < 1.54) = 0.5 - 0.062 = 0.438$$

Because of symmetry

$$P(-1.54 < Z < 0) = 0.438$$

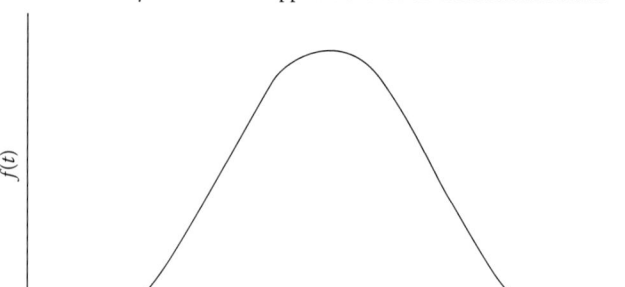

Probability and statistics applications for environmental science

FIGURE 9.1   Normal pdf of time to failure.

and

$$P(Z < -1.54) = 0.062$$

The following probabilities can also be deduced by noting that the area to the right of 1.54 is 0.062 in Figure 9.2.

$$P(-1.54 < Z < 1.54) = 0.876$$

$$P(Z < 1.54) = 0.938$$

$$P(Z > -1.54) = 0.938$$

Table 9.1 can also be used to determine probabilities concerning normal random variables that are not the aforementioned standard normal variables. The required probability is first converted to an equivalent probability about a standard normal variable. For example, if T, the time to failure, is normally distributed with mean $\mu = 200$ and standard deviation $\sigma = 4$, then $(T - 200)/4$ is standard normal variable, and one may write

$$P(T_1 < T < T_2) = P\left(\frac{T_1 - \mu}{\sigma} < \frac{T - \mu}{\sigma} < \frac{T_2 - \mu}{\sigma}\right) \tag{9.2}$$

where

$$\frac{T - \mu}{\sigma} = \frac{T - 200}{4} = Z = \text{standard normal variable} \tag{9.3}$$

TABLE 9.1

Standard Normal Cumulative Probability in Right-Hand Tail (Area under Curve for Specified Values of z_0)

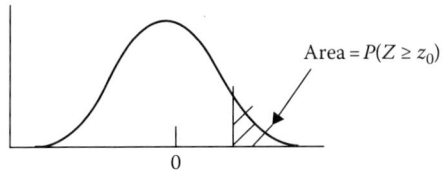

$\text{Area} = P(Z \geq z_0)$

The Standard Normal Distribution

z	0.00	0.01	0.02	0.03	0.04	0.05	0.06	0.07	0.08	0.09
0.0	0.500	0.496	0.492	0.488	0.484	0.480	0.476	0.472	0.468	0.464
0.1	0.460	0.456	0.452	0.448	0.444	0.440	0.436	0.433	0.429	0.425
0.2	0.421	0.417	0.413	0.409	0.405	0.401	0.397	0.394	0.390	0.386
0.3	0.382	0.378	0.374	0.371	0.367	0.363	0.359	0.356	0.352	0.348
0.4	0.345	0.341	0.337	0.334	0.330	0.326	0.323	0.319	0.316	0.312
0.5	0.309	0.305	0.302	0.298	0.295	0.291	0.288	0.284	0.281	0.278
0.6	0.274	0.271	0.268	0.264	0.261	0.258	0.255	0.251	0.248	0.245
0.7	0.242	0.239	0.236	0.233	0.230	0.227	0.224	0.221	0.218	0.215
0.8	0.212	0.209	0.206	0.203	0.200	0.198	0.195	0.192	0.189	0.187
0.9	0.184	0.181	0.179	0.176	0.174	0.171	0.169	0.166	0.164	0.161
1.0	0.159	0.156	0.154	0.152	0.149	0.147	0.145	0.142	0.140	0.138
1.1	0.136	0.133	0.131	0.129	0.127	0.125	0.123	0.121	0.119	0.117
1.2	0.115	0.113	0.111	0.109	0.107	0.106	0.104	0.102	0.100	0.099
1.3	0.097	0.095	0.093	0.092	0.090	0.089	0.087	0.085	0.084	0.082
1.4	0.081	0.079	0.078	0.076	0.075	0.074	0.072	0.071	0.069	0.068
1.5	0.067	0.066	0.064	0.063	0.062	0.061	0.059	0.058	0.057	0.056
1.6	0.055	0.054	0.053	0.052	0.051	0.049	0.048	0.047	0.046	0.046
1.7	0.045	0.044	0.043	0.042	0.041	0.040	0.039	0.038	0.038	0.037
1.8	0.036	0.035	0.034	0.034	0.033	0.032	0.031	0.031	0.030	0.029
1.9	0.029	0.028	0.027	0.027	0.026	0.026	0.025	0.024	0.024	0.023
2.0	0.023	0.022	0.022	0.021	0.021	0.020	0.020	0.019	0.019	0.018
2.1	0.018	0.017	0.017	0.017	0.016	0.016	0.015	0.015	0.015	0.014
2.2	0.014	0.014	0.013	0.013	0.013	0.012	0.012	0.012	0.011	0.011
2.3	0.011	0.010	0.010	0.010	0.010	0.009	0.009	0.009	0.009	0.008
2.4	0.008	0.008	0.008	0.008	0.007	0.007	0.007	0.007	0.007	0.006
2.5	0.006	0.006	0.006	0.006	0.006	0.005	0.005	0.005	0.005	0.005
2.6	0.005	0.005	0.005	0.005	0.005	0.004	0.004	0.004	0.004	0.004
2.7	0.003	0.003	0.003	0.003	0.003	0.003	0.003	0.003	0.003	0.003
2.8	0.003	0.002	0.002	0.002	0.002	0.002	0.002	0.002	0.002	0.002
2.9	0.002	0.002	0.002	0.002	0.002	0.002	0.002	0.002	0.002	0.002

Source: Adapted from Dell, Compare distribution tables, Tulsa, OK. http://www.statsoft.com/textbook/sttable.html.

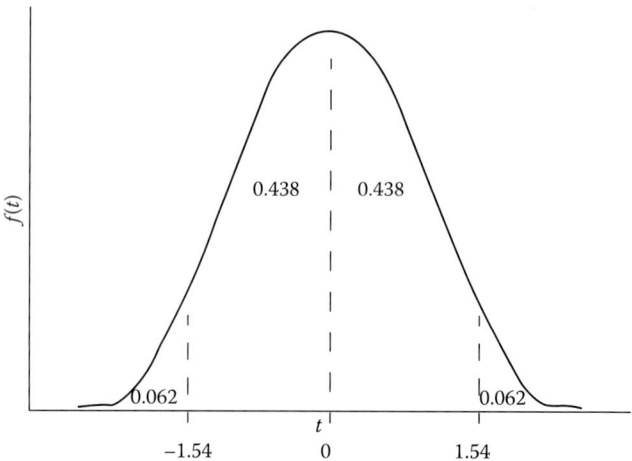

FIGURE 9.2 Areas under a standard normal curve.

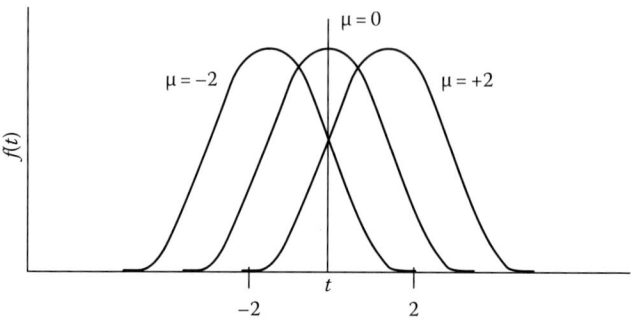

FIGURE 9.3 Normal pdf—varying μ.

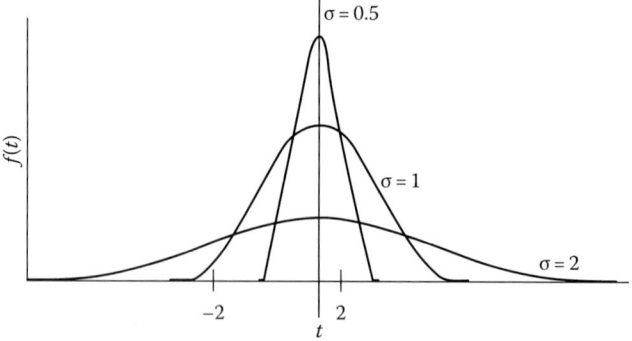

FIGURE 9.4 Normal pdf—varying σ.

Therefore, if $T_1 = 196°F$ and $T_2 = 208°F$ for this example, the describing equation becomes

$$P(196 < T < 208) = P\left(\frac{196 - \mu}{\sigma} < \frac{T - \mu}{\sigma} < \frac{208 - \mu}{\sigma}\right)$$

$$= P\left(\frac{196 - 200}{4} < \frac{T - 200}{4} < \frac{208 - 200}{4}\right)$$

$$= P\left(-1 < \frac{T - 200}{4} < 2\right)$$

$$= P(-1 < Z < 2)$$

$$= 0.341 + 0.477$$

$$= 0.818 = 81.8\%$$

For any random variable X—where X has replaced T—that is normally distributed with mean μ and standard deviation σ, one may now write:

$$P(\mu - \sigma < X < \mu + \sigma) = P\left(-1 < \frac{X - \mu}{\sigma} < 1\right)$$

(9.4)

$$= P(-1 < Z < 1) = 0.68$$

$$P(\mu - 2\sigma < X < \mu + 2\sigma) = P\left(-2 < \frac{X - \mu}{\sigma} < 2\right)$$

(9.5)

$$= P(-2 < Z < 2) = 0.95$$

$$P(\mu - 3\sigma < X < \mu + 3\sigma) = P\left(-3 < \frac{X - \mu}{\sigma} < 3\right)$$

(9.6)

$$= P(-3 < Z < 3) = 0.997$$

This is often referred to as the 68—95—99.7 rule. The rule states that in a *normally* distributed population, 68% of the population falls within one standard deviation of the mean, 95% falls within two standard deviations of the mean, and 99.7% falls within three standard deviations of the mean.

As noted earlier, Table 9.2 provides the area under a standard normal curve between 0 and various positive values of z. $P(0 < Z < 1)$ is the area between 0 and 1. Therefore,

$$P(0 < Z < 1) = 0.3413$$

TABLE 9.2

Areas under a Standard Normal Curve between 0 and z

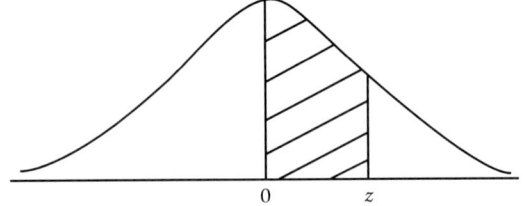

The Standard Normal Distribution

z	0.00	0.01	0.02	0.03	0.04	0.05	0.06	0.07	0.08	0.09
0.0	0.0000	0.0040	0.0080	0.0120	0.0160	0.0199	0.0239	0.0279	0.0319	0.0359
0.1	0.0398	0.0438	0.0478	0.0517	0.0557	0.0596	0.0636	0.0675	0.0714	0.0753
0.2	0.0793	0.0832	0.0871	0.0910	0.0948	0.0987	0.1026	0.1064	0.1103	0.1141
0.3	0.1179	0.1217	0.1255	0.1293	0.1331	0.1368	0.1406	0.1443	0.1480	0.1517
0.4	0.1554	0.1591	0.1628	0.1664	0.1700	0.1736	0.1772	0.1808	0.1844	0.1879
0.5	0.1915	0.1950	0.1985	0.2019	0.2054	0.2088	0.2123	0.2157	0.2190	0.2224
0.6	0.2257	0.2291	0.2324	0.2357	0.2389	0.2422	0.2454	0.2486	0.2517	0.2549
0.7	0.2580	0.2611	0.2642	0.2673	0.2704	0.2734	0.2764	0.2794	0.2823	0.2852
0.8	0.2881	0.2910	0.2939	0.2967	0.2995	0.3023	0.3051	0.3078	0.3106	0.3133
0.9	0.3159	0.3186	0.3212	0.3238	0.3264	0.3289	0.3315	0.3340	0.3365	0.3389
1.0	0.3413	0.3438	0.3461	0.3485	0.3508	0.3531	0.3554	0.3577	0.3599	0.3621
1.1	0.3643	0.3665	0.3686	0.3708	0.3729	0.3749	0.3770	0.3790	0.3810	0.3830
1.2	0.3849	0.3869	0.3888	0.3907	0.3925	0.3944	0.3962	0.3980	0.3997	0.4015
1.3	0.4032	0.4049	0.4066	0.4082	0.4099	0.4115	0.4131	0.4147	0.4162	0.4177
1.4	0.4192	0.4207	0.4222	0.4236	0.4251	0.4265	0.4279	0.4292	0.4306	0.4319
1.5	0.4332	0.4345	0.4357	0.4370	0.4382	0.4394	0.4406	0.4418	0.4429	0.4441
1.6	0.4452	0.4463	0.4474	0.4484	0.4495	0.4505	0.4515	0.4525	0.4535	0.4545
1.7	0.4554	0.4564	0.4573	0.4582	0.4591	0.4599	0.4608	0.4616	0.4625	0.4633
1.8	0.4641	0.4649	0.4656	0.4664	0.4671	0.4678	0.4686	0.4693	0.4699	0.4706
1.9	0.4713	0.4719	0.4726	0.4732	0.4738	0.4744	0.4750	0.4756	0.4761	0.4767
2.0	0.4772	0.4778	0.4783	0.4788	0.4793	0.4798	0.4803	0.4808	0.4812	0.4817
2.1	0.4821	0.4826	0.4830	0.4834	0.4838	0.4842	0.4846	0.4850	0.4854	0.4857
2.2	0.4861	0.4864	0.4868	0.4871	0.4875	0.4878	0.4881	0.4884	0.4887	0.4890
2.3	0.4893	0.4896	0.4898	0.4901	0.4904	0.4906	0.4909	0.4911	0.4913	0.4916
2.4	0.4918	0.4920	0.4922	0.4925	0.4927	0.4929	0.4931	0.4932	0.4934	0.4936
2.5	0.4938	0.4940	0.4941	0.4943	0.4945	0.4946	0.4948	0.4949	0.4951	0.4952
2.6	0.4953	0.4955	0.4956	0.4957	0.4959	0.4960	0.4961	0.4962	0.4963	0.4964
2.7	0.4965	0.4966	0.4967	0.4968	0.4969	0.4970	0.4971	0.4972	0.4973	0.4974
2.8	0.4974	0.4975	0.4976	0.4977	0.4977	0.4978	0.4979	0.4979	0.4980	0.4981
2.9	0.4981	0.4982	0.4982	0.4983	0.4984	0.4984	0.4985	0.4985	0.4986	0.4986
3.0	0.4987	0.4987	0.4987	0.4988	0.4988	0.4989	0.4989	0.4989	0.4990	0.4990

Source: Adapted from Dell, Compare distribution tables, Tulsa, OK. http://www.statsoft.com/textbook/sttable.html.

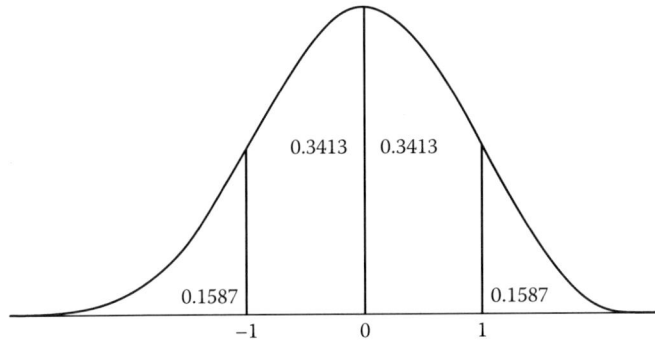

FIGURE 9.5   Normal areas.

Note once again that because the standard normal curve is symmetric about 0, the area to the right of 0 is equal to the area to the left of 0, namely, 0.5. Because the area over (0, 1) is 0.3413, the area over (1, ∞) is 0.5 − 0.3413=0.1587. By symmetry, the area over (−1, 0) is 0.3413, and the area over (−∞,−1) is 0.1587. These areas are bounded by ordinates erected at −1, 0, and 1, as noted in Figure 9.5.

Additional probabilities are represented by areas under the standard normal curve over certain intervals. From Figure 9.5, one obtains the following additional required probabilities:

1. $P(-1 < Z < 0)$ = Area over (−1, 0) = 0.3413
2. $P(-1 < Z < 1)$ = Area over (−1, 1) = 0.6826
3. $P(Z > -1)$ = Area over (1, ∞) = 0.3413 + 0.5 = 0.8413
4. $P(Z < 1)$ = Area over (−∞, 1) = 0.5 + 0.3413 = 0.8413
5. $P(Z > 1)$ = Area over (1, ∞) = 0.1587

The details of a probability calculation for nonstandard normal variable is presented in the following example. The measurement of the pitch diameter of the threat of a fitting is normally distributed with mean 0.4008 in. and standard deviation 0.0004 in. The specifications are given as 0.4000 ± 0.001 in. The thread diameter meets specifications if the diameter lies between 0.4000 − 0.001 and 0.4000 + 0.001 in. If the thread diameter does not meet specifications, it is considered defective. What is the probability that a defective thread will occur? Note that in this case, the random variable is not a standard normal variable but is normally distributed. The random variable is the pitch diameter X of the thread of a fitting selected at random from a manufactured output. The required probability is the probability that a defect occurs, that is, the thread of a fitting produced lies outside the given specification. Therefore, the required probability is the probability that X exceeds 0.4000 + 0.001 or is less than 0.4000 − 0.001 in. This required probability is first converted into a probability about a standard normal variable by making use of the fact that if X is normally distributed with mean μ and standard deviation σ and that $Z = (X - \mu)/\sigma$ is a standard normal variable,

$$P(X > 0.4010) + P(X < 0.3990)$$

$$= 1 - P(0.3990 < X < 0.4010)$$

$$= 1 - P\left[\frac{0.3990 - 0.4008}{0.0004} < \frac{-0.4008}{-0.0004} < \frac{0.4010 - 0.4008}{-0.0004}\right]$$

$$= 1 \quad P(4.5 < Z < 0.5)$$

Here, $P(-4.5 < Z < 0.5)$ is the area under a standard normal curve between −4.5 and 0.5. Now proceed to find the area between 0 and 0.5 either from Table 9.1 or Table 9.2 and then the area between −4.5 and 0. The area between 0 and 0.5 is 0.1915. The area between −4.5 and 0 is the same area between 0 and 4.5. The latter area is approximately 0.5. Therefore, the total area between −4.5 and 0.5 is 0.6915, so that

$$P(-4.5 < Z < 0.5) = 0.6915$$

and

$$1 - P(-4.5 < Z < 0.5) = 0.3085$$

Thus, the probability that a defective thread occurs is 0.3085, or approximately 31%.

The probabilities given in Equations 9.4 through 9.6 are the key sources of the percentages cited earlier. These can be used to interpret the standard deviation s of a *sample* of observations on a normal random variable, as a measure of dispersion about the *sample* mean X.

The normal distribution is used to obtain probabilities concerning the mean X of a sample of n observations on a random variable X, if X is normally distributed with mean μ and standard deviation $\sigma/\sqrt{n}$. For example, suppose X is normally distributed with mean 100 and standard deviation 2. Then, $\bar{X}$, the mean of a sample of 16 observations on X, is normally distributed with mean 100 and standard deviation 0.5. To calculate the probability that X is greater than 101, one would write

$$P(\bar{X} > 101) = P\left[\frac{\bar{X} - 100}{0.5} > \frac{101 - 100}{0.5}\right]$$

$$P(\bar{X} > 101) = P\left[\frac{\bar{X} - 100}{0.5} > \frac{101 - 100}{0.5}\right]; \quad Z = \frac{\bar{X} - 100}{0.5}$$

$$= P(Z > 2)$$

$$= 0.023 = 2.3\%$$

If X is *not* normally distributed, then $\bar{X}$, the mean of a sample of n observations on X, is *approximately* normally distributed with mean μ and standard deviation $\sigma/\sqrt{n}$, provided the sample size n is large (>30). This result is based on an important theorem in probability called the *central limit theorem*.

Suppose, for example, that the pdf of the random variable X is specified by

$$f(x) = 1/2; \quad 0 < x < 2$$

$$= 0; \quad \text{elsewhere}$$

Application of the equation defining the expected value of X (see Chapter 4) gives

$$\mu = E(X) = \int_0^2 x \left(\frac{1}{2} \right) dx = 1$$

$$E(X^2) = \int_0^2 x^2 \left(\frac{1}{2} \right) dx = \frac{4}{3}$$

Therefore, application of the defining equation for the variance (see also Chapter 4) yields

$$\sigma^2 = E(X^2) - \mu^2 = \frac{4}{3} - 1 = \frac{1}{3}$$

Thus, if $\bar{X}$ is the mean of a random sample of 48 observations on X, $\bar{X}$ is approximately normally distributed with mean 1 and standard deviation $\sigma/\sqrt{n}$. The latter term is therefore given by

$$\frac{\sigma}{\sqrt{n}} = \frac{\sqrt{1/3}}{\sqrt{48}} = \sqrt{\frac{1}{(3)(48)}} = \sqrt{\frac{1}{144}}$$

$$= \frac{1}{12}$$

The following example is now provided:

$$P(\bar{X} > 9/8) = P\left[\frac{\bar{X} - 1}{1/12} > \frac{9/8 - 1}{1/12} \right]$$

$$= P(Z > 1.5)$$

$$= 0.067 = 6.7\%$$

One of the principal applications of the normal distribution in reliability calculations and hazard risk analysis is the distribution of time to failure due to *wear out*. Suppose, for example, that a production lot of thermometers, to be employed in an incinerator especially designed to withstand high temperatures and intense vibrations, has just come off the assembly line. A sample of 25 thermometers from the lot is tested under the specified heat and vibration conditions. Time to failure, in hours, is recorded for each of the 25 thermometers. Application of the equations for sample mean $\bar{X}$ and sample variance s^2 yields

$$\bar{X} = 125; \quad s^2 = 92$$

Past experience indicates that the *wear-out* time of this *unit*, like that of a large variety of products in many different industries, tends to be normally distributed. Using the above calculated values of $\bar{X}$ and s as best estimates of μ and σ, one may obtain the 110 h reliability of the thermometers.

$$P(\bar{X} > 110) = P\left[\frac{\bar{X} - 125}{\sqrt{92}} > \frac{110 - 125}{\sqrt{92}}\right]$$

$$= P(Z > -1.56)$$

$$= 0.94 = 94\%$$

As indicated earlier, the normal distribution is symmetric. The data from a normal distribution could be plotted on special graph paper, known as the normal probability paper. The resulting plot appears as a straight line. The parameters μ and σ can be estimated from such a plot. This procedure is demonstrated in the work of Theodore and Taylor.[1] A nonlinear plot on this paper is indicative of nonnormality.

Interestingly, the normal approximation to the binomial distribution has received considerable attention in the past in the probability/statistics literature. The approximation to the binomial distribution simply involves replacing the mean (μ) and variance (σ^2)—in the standard normal variable Z—by np and $np(1 - p)$, respectively.

Actual (experimental) data have shown many physical variables to be normally distributed. Examples include physical measurements on living organisms, molecular velocities in an ideal gas, scores on an intelligence test, the average temperatures in a locality, height of men belonging to a certain race, experimental measurement subject to random errors, and time for a delivery truck to travel alone a particular route. Other variables, though not normally distributed per se, sometimes approximate a normal distribution after an appropriate transformation. Examples include taking the logarithm (see Chapter 10) or square root of the original variable. The normal distribution also has the advantage that it is tractable mathematically. Consequently, many of the techniques of statistical inference have been derived under the assumption of underlying normal variants.

On account of the prominence (and perhaps the name) of the normal distribution, it is sometimes assumed that a random variable is normally distributed, unless proven otherwise. This notion could lead to incorrect results. For example, the normal distribution is generally inappropriate as a model of time to failure. Frequently, a normal distribution provides a reasonable approximation to the main part of this distribution, but is inadequate at one or both tails. Finally, certain phenomena are just not symmetrically distributed, as is required for normality. The errors of incorrectly assuming normality depend on the use to which this assumption is applied. Many statistical models and methods derived under this assumption remain valid under moderate deviations from it. On the other hand, if the normality assumption is used to determine the proportion of *items* above or below some extreme limit (e.g., at the tail of the distribution), serious errors might result.

In addition, if

$$F(z) = \int_{0}^{z} \frac{1}{\sqrt{2\pi}} e^{\frac{z^2}{2}} dz \qquad (9.7)$$

the term

$$1 - 2\left[1 - F(z)\right] \qquad (9.8)$$

is referred to as the *confidence level* and provides the probability that the *result* will be in the range $\mu = \pm z$. The term $1 - 2[1 - F(z)]$ is often denoted by α, i.e.,

$$\alpha = 1 - 2\left[1 - F(z)\right] \qquad (9.9)$$

and referred to as the *level of significance*.[1,2] The term α therefore represents the probability of being in error if one observes a result that is displaced from the mean by $z\sigma$. Any detailed discussion of this is beyond the scope of this book. The reader is referred to the literature for details and illustrative examples.[1,2]

The *reliability* function corresponding to the normally distributed failure time is given by

$$R(t) = \frac{1}{\sqrt{2\pi}\sigma} \int_{t}^{\infty} \exp\left[-\frac{1}{2}\left(\frac{t-\mu}{\sigma}\right)^2\right] dt \qquad (9.10)$$

The corresponding failure rate is obtained by the substitution of Equation 9.10, which states that the failure rate $Z(t)$ is related to the reliability $R(t)$ by Equation 9.11.[1,2]

$$Z(t) = \frac{R'(t)}{R(t)} \qquad (9.11)$$

since

$$R'(t) = F'(t) = \frac{d\left[f(t)\right]}{dt} \tag{9.12}$$

And, substituting $f(t)$ from Equation 9.1 and $R(t)$ from Equation 9.10 yields

$$Z(t) = \frac{\exp\left[-\frac{1}{2}\left(\frac{t-\mu}{\sigma}\right)^2\right]}{\displaystyle\int_t^\infty \exp\left[-\frac{1}{2}\left(\frac{t-\mu}{\sigma}\right)^2\right]dt} \tag{9.13}$$

as the failure rate corresponding to a normally distributed failure time.

Summarizing, one reason the normal distribution is so important is that a number of natural phenomena are normally distributed or closely approximate it. In fact, many experiments, when repeated a large number of times, will approach the normal distribution curve. In its pure form, the normal curve is a continuous, symmetrical, smooth curve shaped like the one shown earlier. Naturally, a finite distribution of discrete data can only approximate this curve.

The normal curve has the following definite relations to the descriptive measures of a distribution. The normal distribution curve is symmetrical; therefore, both the mean and the median are always found to be in the middle of the curve. Recall that, in general, the mean and median of an asymmetrical distribution do not coincide. The normal curve ranges along the x axis from minus infinity to plus infinity. Therefore, the range of a normal distribution is infinite. The standard deviation, σ, becomes the most meaningful measure when related to the normal curve. A total of 68.2% of the area lying under a normal curve is included by the part ranging from one standard deviation below to one standard deviation above the mean. A total of 95.4% lies −2 to +2 standard deviations from the mean. By using the tables presented earlier and those found in standard statistics texts and handbooks, one can determine the area lying under any part of the normal curve.

This section concludes by examining some topics that Shaefer and Theodore[2] have classified as contemporary statistics. The following subject areas are briefly reviewed:

1. Confidence intervals for means
2. Confidence intervals for proportions
3. Hypothesis testing
4. Hypothesis test for means and proportions

Shaefer and Theodore[2] provide numerous illustrative examples associated with these topics. These four topics will receive additional treatment in Chapters 12 and 13.

CONFIDENCE INTERVALS FOR MEANS

As described in an earlier section, the sample mean $\overline{X}$ constitutes a so-called point estimate of the mean μ of the population from which the same was selected at random. Instead of a *point* estimate, an *interval* estimate of μ may be required along with an indication of the confidence that can be associated with the interval estimate. Such an interval estimate is called a *confidence interval*, and the associated confidence is indicated by a *confidence coefficient*. The length of the confidence interval varies directly with the confidence coefficient for fixed values of the sample size n; the larger the value of n, the shorter the confidence interval. Thus, for fixed values of the confidence coefficient, the limits that contain a parameter with a probability of 95% (or some other stated percentage) are defined as 95% (or any other percentage) confidence limits for the parameter; the interval between the confidence limit is the aforementioned confidence interval.

For normal distributions, the confidence coefficient Z can be obtained from the standard normal table for various confidence limits or corresponding levels of significance for μ. This can be employed to obtain the probability of a value falling inside or outside the range $\mu \pm Z\sigma$. For example, when $Z = 2.58$, one can say that the level of significance is 1%. The statistical interpretation of this is as follows. If an observation deviates from the mean by at least $\pm 2.58\sigma$, the observation is significantly *different* from the body of data on which the describing normal distribution is based. Further, the probability that this statement is in error is 1%, that is, the conclusion drawn from a rejected observation will be wrong 1% of the time.

One can also state that there is a 99% probability that an observation will fall *within* the range $\mu \pm 2.58$. This degree of confidence is referred to as 99% *confidence level*. The limits, $\mu - 2.58\sigma$ and $\mu + 2.58\sigma$, are defined as the *confidence limits*, whereas the difference between the two values is defined as the *confidence interval*. Once again, this essentially states that the actual (or true) mean lies within the interval $\mu - 2.58\sigma$ and $\mu + 2.58\sigma$ with a 99% probability of being correct. The foregoing analysis can be extended to provide the difference of two population *means*, that is, $X_2 - X_1$. For this case, the confidence limits are

$$\overline{X_2} - \overline{X_1} \pm Z \sqrt{\frac{\sigma_2^2}{n_2} + \frac{\sigma_1^2}{n_1}} \tag{9.14}$$

This section also serves to introduce Student's t distribution (named after a student statistician with the pen name of W. Gosset). It is common to use this distribution if the sample size is small. For a random sample of size n selected from a normal population, the term $(\overline{X} - \mu)/(s/\sqrt{n})$ has a Student's distribution with $(n - 1)$ degrees of freedom.[2] *Degree of freedom* is a label used for the parameter appearing in Student's distribution pdf. Details are provided by Shaefer and Theodore.[2] An additional review is presented in Chapter 13.

CONFIDENCE INTERVALS FOR PROPORTIONS

Consider a random variable X that has a binomial distribution. For large n, the random variable X/n is approximately normally distributed with mean p and standard deviation equal to the square root of pq/n. If n is the size of a random sample from a population in which p is the proportion of elements classified as *success* because of the possession of a specified characteristic, it serves as the basis for constructing a confidence interval for P, the corresponding population proportion. The term n is *large* means that np and $n(1 - p)$ are both greater than 5.

One can generate an inequality whose probability equals the desired confidence coefficient using the large sample distribution of the sample proportion. Note, once again, that X/n, the sample proportion, is approximately normally distributed with mean p and standard deviation equal to the square root of pq/n. Therefore, $(X/n-p)/(pq/n)^{1/2}$ is *approximately* a standard normal variable and

$$P\left(-Z < \frac{(X/n)-p}{\sqrt{pq/n}} < Z\right) = \text{confidence interal; fraction basis} \qquad (9.15)$$

where Z is obtained from the table of the standard normal distribution as the value such that the area over the interval from $-Z$ to Z is the stated fractional basis. This inequality can then be converted into a statement about P:

$$\frac{X}{n} - Z\sqrt{\frac{pq}{n}} < P < \frac{X}{n} + Z\sqrt{\frac{pq}{n}} \qquad (9.16)$$

The endpoints of the statement may then be evaluated after replacing p and q in the endpoints by the observed value of X/n and $1 - X/n$, respectively.

The large sample distribution of the sample proportion X/n also provides the basis for determining the sample size n for estimating the population proportion p with maximum allowable error E and specified confidence.

HYPOTHESIS TESTING

In *hypothesis testing*, one can make a statement (hypothesis) about an unknown population parameter and then use statistical methods to determine (test) whether the observed sample data support that statement. Note that a statistical hypothesis is an assumption made about some parameter, that is, about a statistical measure of a population. One may also define *hypothesis* as a statement about one or more parameters. Synonyms for hypothesis could include *assumption* or *guess*. A hypothesis can be tested to verify its credibility.

An important notion to understand is the relationship of the null hypothesis to the *alternative hypothesis* (or hypotheses). In all applications, the *null hypothesis* and alternatives are written in terms of population parameters. For example, if μ represents the population mean (e.g., temperature), one could write

Null hypothesis is H_0 : $\mu = 100$

Alternative hypothesis is H_1 : $\mu \neq 100$

Since statistics are only estimates of a *true* population parameter based on a sample of observations, it is reasonable to expect that the value of the estimate will deviate from the value of the true population parameter. The method called *hypothesis testing* is used to decide whether the value of the statistic is consistent with a hypothesized value from the population parameter. When an estimate is made, a hypothesis is tested concerning an assumed population parameter.

In hypothesis testing, the statistician or engineer or applied scientist use various techniques to decide whether to accept or reject hypotheses. If, for example, someone assumed a population mean of 70 (e.g., a thermometer reading), and a sample from that population were selected that had a mean of 80, one might want to perform a test or calculation to see if the population mean of 70 could still be reasonably assumed. The individual would then employ a statistical technique to either accept or reject the hypothesis that the population mean equals 70.

In the above example, a sample mean equal to 80 is used to test a hypothesized population mean of 70. The notation employed is as follows:

$$\bar{X} = 80 \text{ (sample mean)}$$

and the hypothesis is

$$\mu = 70 \left(\text{population mean} \right)$$

As indicated earlier, each null hypothesis is accompanied by an alternate hypothesis.

HYPOTHESIS TEST FOR MEANS AND PROPORTIONS

As described in the previous section, testing a statistical hypothesis concerning a population mean involves the setting of a rule for deciding, on the basis of a random sample from the population, whether to reject the hypothesis being tested; this is the so-called aforementioned null hypothesis. The test is formulated in terms of a test statistic, that is, some function of the sample observations. In the case of a large sample ($n > 30$), testing a hypothesis concerning the population mean μ involves one of the following test statistics:

$$Z = \frac{\bar{X} - \mu_0}{\sigma/n} \tag{9.17}$$

where

$\bar{X}$ is the sample mean
μ_0 is the value of μ specified by null hypothesis H_0
σ is the population standard deviation
n is the sample size

When σ is unknown, the sample standard deviations may be employed to estimate it. This test statistic is approximately distributed as a standard normal variable when the sample originates from a normal population whose standard deviation σ is known.

TABLE 9.3

Normal Distributions: Critical Regions

Alternative Hypothesis	Critical Region
$\mu < \mu_0$ (one tail)	$Z < -Z_\alpha$
$\mu > \mu_0$ (one tail)	$Z > Z_\alpha$
$\mu \neq \mu_0$ (two tail)	$Z < Z_{\alpha/2}$ or $Z > Z_{\alpha/2}$

The values of the test statistic for which the null hypothesis H_0 is rejected constitute the *critical region*. The critical region depends on the alternative hypothesis H_1 and the tolerated region for three alternative hypotheses (see Table 9.1). The term Z_α is the value that a standard normal variable exceeds with probability α, and $Z_{\alpha/2}$ is defined similarly. Typical values for the tolerated probability of a type I error[2] are 0.05 and 0.01, with 0.05 the more common selection. These are summarized in Table 9.3.

ILLUSTRATIVE EXAMPLES

Illustrative Example 9.1

Assume that machined bolt sizes have a normal distribution with mean $\mu = 100$ mm and standard deviation $\sigma = 20$ mm. Find the percentage of bolts whose sizes fall between

1. 80 and 120 mm
2. 40 and 160 mm
3. Over 160

Solution

All the bolts are units of the standard deviation $\sigma = 20$ mm from the mean $\mu = 100$ mm. One can apply the aforementioned 68–95–99.7 rule or Table 9.1 to obtain

1. 68%
2. 99.7%
3. $(0.5)(100 - 99.7) = 0.15\%$

Illustrative Example 9.2

The temperature T of a cooling water source is normally distributed with mean $\mu = 68°F$ and standard deviation $\sigma = 6°F$. Find the probability that the temperature of the cooling water is

1. Between 70°F and 80°F
2. Less than 60°F

Solution

First convert the temperature values into Z values in standard units and refer to Table 9.1.

For (1)

$$70°F \text{ in standard normal units} = (70-68)/6 = 0.33$$
$$80°F \text{ in standard normal units} = (80-68)/6 = 2.00$$

Therefore,

$$P(70°F \leq T \leq 80°F) = P(0.33 \leq Z \leq 2.00)$$
$$= 0.4722 - 0.1293$$
$$= 0.3479$$

For (2)

$$60°F \text{ in standard normal units} = (60-68)/6 = -1.33$$

This is a one-sided probability with $-1.33 < 0$. Using Table 9.1. symmetry, and noting that half the area under the curve is 0.500, one obtains

$$P(T \leq 60°F) = P(Z \leq -1.33) = P(Z \geq 1.33)$$
$$= 0.5000 - 0.4082$$
$$= 0.0918$$

Illustrative Example 9.3

The mass M of 800 chemical batches produced in an industrial operation is normally distributed with mean $\mu = 140$ lb and standard deviation $\sigma = 10$ lb. Estimate the number of batches N with mass M

1. Between 138 and 148 lb
2. More than 152 lb

Solution

Once again, convert the M values into Z values in standard normal units and apply Table 9.1.

For (1)

$$138 \text{ in standard units} = (138-140)/10 = -0.2$$
$$148 \text{ in standard units} = (148-140)/10 = 0.8$$

The area between $-0.2 < 0 < 0.8$ is required. Therefore, from Table 9.1,

$$P(138 \leq M \leq 148) = P(-0.2 \leq Z \leq 0.8)$$

$$= 0.2881 + 0.0793$$

$$= 0.3674$$

Thus,

$$N = 800(0.3674) \approx 300$$

For (2)

$$152 \text{ in standard units} = (152 - 140)/10 = 1.20$$

This is a one-sided application requiring the area for $0 < 1.20$. Using Table 9.1 and noting once again that half the area under the curve is 0.5000, one obtains

$$P(M \geq 152) = P(Z \geq 1.2) = 0.5000 - 0.3849$$

$$= 0.1151$$

Thus,

$$N = 800(0.1151) \approx 92$$

Illustrative Example 9.4

The parts-per-million concentration of a particular toxic substance in a waste-water stream is known to be normally distributed with a mean $\mu = 100$ $\mu g/m^3$ and a standard deviation $\sigma = 2.0$ $\mu g/m^3$. Calculate the probability that the toxic concentration, C, is between 98 and 104 $\mu g/m^3$.

Solution
Because C is normally distributed with $\mu = 100$ and a standard deviation $\sigma = 2.0$, then $(C - 100)/2$ is a standard normal variable and

$$P(98 < C < 104 = P\left(-1 < \left[\frac{C-100}{2}\right] < 2\right)$$

$$= P(-1 < Z < 2)$$

From the values in the standard normal table (refer to Table 9.1),

$$P(98 < C < 104) = 0.341 + 0.477$$

$$= 0.818 = 81.8\%$$

Illustrative Example 9.5

Acceptance limits require the plate-to-plate spacing in an electrostatic precipitator (ESP) to be between 24 and 25 cm. Spacing from other installations suggests that it is normally distributed with an average length of 24.6 cm and a standard deviation of 0.4 cm. What proportion of the plate spacing in a typical unit can be assumed unacceptable?

Solution

The problem is equivalent to determining the probability that an observation or sample from a normal distribution with parameters $\mu = 0$ and $\sigma = 1$ either exceeds

$$Z = \frac{25.0 - 24.6}{0.4} = 1$$

or falls below

$$Z = \frac{24.0 - 24.6}{0.4} = -1.5$$

and the corresponding probabilities are 0.159 and 0.067, respectively, from Table 9.1. Consequently, the proportion of unacceptable spacing is

$$P(\text{unacceptable spacing}) = P(Z > 1) + P(Z < -1.5)$$
$$= 0.159 + 0.067$$
$$= 0.226$$
$$= 22.6\%$$

Illustrative Example 9.6

The temperature of a polluted estuary during the summer months is normally distributed with a mean of 56°F and a standard deviation of 3.0°F. Calculate the probability that the temperature is between 55°F and 62°F.

Solution

Normalizing the temperature T gives

$$Z_1 = \frac{55 - 56}{3.0} = -0.333$$

$$Z_2 = \frac{60 - 56}{3.0} = 2.0$$

Thus,

$$P(0.333 < Z < 2.0) = P(0.0 < Z < 2.0) - P(0.0 < Z < 0.333)$$

$$= 0.4722 - 0.1293$$

$$= 0.6015$$

$$= 61.15\%$$

Illustrative Example 9.7

The regulatory specification on a toxic substance in a solid waste ash calls for a concentration level of 1.0 ppm or less. Earlier observations of the concentration of the ash, C, indicate a normal distribution with a mean of 0.60 ppm and a standard deviation of 0.20 ppm. Estimate the probability that ash will exceed the regulatory limit.

Solution

This problem requires the calculation of $P(C > 1.0)$. Normalizing the variable C,

$$P\left(\left[\frac{C - 0.6}{0.2}\right] > \left[\frac{1.0 - 0.6}{0.2}\right]\right)$$

$$P(Z > 2.0)$$

From the standard normal table (Table 9.1),

$$P(Z > 2.0) = 0.0228$$

$$= 2.28\%$$

For this situation, the area to the right of the 2.0 is 2.28% of the total area. This represents the probability that ash will exceed the regulatory limit of 1.0 ppm.

Illustrative Example 9.8

The diameter D of wires installed in an ESP has a standard deviation of 0.01 in. At what value should the mean be if the probability of its exceeding 0.21 in. is to be 1%?

Solution

For this one-tailed case (with an area of 0.01),

$$P\left(\left[\frac{D - \mu}{\sigma}\right] > \left[\frac{0.21 - \mu}{\sigma}\right]\right) = 0.01$$

where

$$Z = \frac{D - \mu}{\sigma}$$

Therefore,

$$P\left(Z > \left[\frac{0.21 - \mu}{\sigma}\right]\right) = 0.01$$

For a one-tailed test at the 1% (0.01) level,

$$Z = 2.326$$

so that

$$2.326 = \frac{0.21 - \mu}{0.01}$$

$$\mu = 0.187 \text{ in}$$

Illustrative Example 9.9

The lifetime, T, of a circuit board employed in an incinerator is normally distributed with a mean of 2500 days. What is the largest lifetime variance the installed circuit boards can have if 95% of them need to last at least 365 days?

Solution
For this application,

$$P(T > 365) = 0.95$$

Normalizing gives

$$P\left(\left[\frac{T - \mu}{\sigma}\right] > \left[\frac{365 - 2500}{\sigma}\right]\right) = 0.95$$

$$P\left(Z > -\frac{2135}{\sigma}\right) = -1.645$$

The following equation must apply for this condition:

$$-\frac{2135}{\sigma} = -1.645$$

Solving,

$$\sigma = 1298 \text{ days} = 1298 \text{ d}$$

In addition,

$$\sigma^2 = 1.68 \times 10^6 \text{ d}^2$$

Illustrative Example 9.10

Let X denote the coded quality of a bag fabric used in a particular utility baghouse. Assume that X is normally distributed with a mean of 10 and a standard deviation of 2. Find c such that $P(|X - 10| < c) = 0.90$.

Solution

Because X is normally distributed with a mean of 10 and a standard deviation of 2, $(X - 10)/2$ is a standard normal variable. For this two-sided test,

$$P(|X - 10| < c) = P(-c < (X - 10) < c)$$

$$= P\left(\frac{-c}{2} < \frac{X - 10}{2} < \frac{c}{2}\right)$$

$$= P\left(\frac{-c}{2} < Z < \frac{c}{2}\right)$$

$$= 2P\left(0 < Z < \frac{c}{2}\right); \quad \text{due to symmetry}$$

Because

$$P(|X - 10| < c) = 0.90$$

$$2P\left(0 < Z < \frac{c}{2}\right) = 0.90$$

$$P\left(0 < Z < \frac{c}{2}\right) = 0.45$$

For this condition to apply,

$$\frac{c}{2} = 1.645$$

$$c = 3.29$$

Illustrative Example 9.11

In the random samples of 16 observations, where the temperature T is normally distributed with a mean of 101 and a standard deviation of 4, find $P(\bar{X} > 103)$.

Solution

If $\bar{X}$ is the mean of a sample from a normal population with a mean μ and a standard deviation σ, then $\bar{X}$ (as noted in an earlier section) is normally distributed with mean μ and standard deviation $\sigma/\sqrt{n}$. If the population sampled is not normal, then $\bar{X}$ is approximately normally distributed with mean μ and standard deviation $\sigma/\sqrt{n}$, provided the sample size n is large, that is, $n > 30$. This sample distribution of the sample mean is based on the central limit theorem discussed earlier.

Note that $\bar{X}$ is normally distributed with a mean of 101 and a standard deviation of $4/\sqrt{16} = 1.0$. Therefore,

$$P\left(\bar{X} > 101\right) = P\left[\frac{\bar{X} - 101}{1.0} > \frac{103 - 101}{1.0}\right]$$

$$= P\left(Z > 2\right)$$

$$= 0.023 = 2.3\%$$

REFERENCES

1. L. Theodore and F. Taylor, *Probability and Statistics*, Theodore Tutorials (originally published by USEPA, RTP, NC), East Williston, NY, 1996.
2. S. Shaefer and L. Theodore, *Probability and Statistics Applications in Environmental Science*, CRC Press/Taylor & Francis Group, Boca Raton, FL, 2007.
3. http://www.statsoft.com/textbook/sttable.html.
4. L. Theodore, personal notes, East Williston, NY, 2013.

10 Log-Normal Distribution

INTRODUCTION

Before proceeding to the presentation on log-normal distribution, the interests of the reader would be better served with several short introductory paragraphs on logarithms. The logarithm of a number is the exponent of that power to which another number, the base, must be raised to give the number first named. Any positive number greater than 1 might serve as a base. Two have been selected by the technical community. One base, 2.718 denoted by the letter e, gives rise to a system of logarithms that are conveniently applied in engineering and science. These are referred to as Napierian or hyperbolic logarithms. The other base used is 10, giving logarithms particularly adapted for use in computation, referred to by some as common or Briggsian logarithms. Tables of logarithms given without designation invariable refer to base 10. Since most numbers are irrational powers of 10, a common logarithm, in general, consists of an integer, which is called the characteristic and an endless decimal, the mantissa.[1,2]

The integration of some log terms arise in practice. Ten of these integrations with limits of 0–1 are provided in the following.

$$\int_0^1 (\log x)^n dx = (-1)^n \cdot n!$$

$$\int_0^1 \left(\log \frac{1}{x}\right)^{1/2} dx = \frac{\sqrt{\pi}}{2}$$

$$\int_0^1 \left(\log \frac{1}{x}\right)^{-\frac{1}{2}} dx = \sqrt{\pi}$$

$$\int_0^1 \left(\log \frac{1}{x}\right)^n dx = n!$$

$$\int_0^1 x \log(1-x) dx = -\frac{3}{4}$$

$$\int_0^1 x \log(1+x) dx = \frac{1}{4}$$

$$\int_0^1 \frac{\log x}{1+x} dx = -\frac{\pi^2}{12}$$

$$\int_0^1 \frac{\log x}{1-x} dx = -\frac{\pi^2}{6}$$

$$\int_0^1 \frac{\log x}{1-x^2} dx = -\frac{\pi^2}{8}$$

$$\int_0^1 \log\left(\frac{1+x}{1-x}\right) \frac{dx}{x} = \frac{\pi^2}{4}$$

The use of logarithmic scales on graphs also finds applications. Log–log coordinates are useful in plotting equations of the frequently occurring form of $y = bx^a$. If logarithms (natural logarithms also apply) are taken on both sides of this equation, one obtains $\log y = a \log x + \log b$. If $\log y$ is plotted as a function of $\log x$, or equivalently, if y is plotted as a function of x on log–log coordinates, a straight line of slope a and intercept $\log b$ will result for log-normal distributions.

A nonnegative random variable X has a log-normal distribution whenever $\ln X$, that is, the natural logarithm of X, has a normal distribution. The probability distribution function (pdf) of a random variable X having a log-normal distribution is specified by

$$f(x) = \frac{1}{\sqrt{2\pi}\beta} x^{-1} \exp\left[-\frac{(\ln x - \alpha)^2}{2\beta^2}\right]; \quad x > 0 \tag{10.1}$$

$$= 0; \quad \text{elsewhere}$$

The mean and variance of a random variable X having a log-normal distribution are given by

$$\mu = e^{\alpha + \beta^2/2} \tag{10.2}$$

$$\sigma^2 = e^{2\alpha + \beta^2}\left(e^{\beta^2} - 1\right) \tag{10.3}$$

Figure 10.1 plots the pdf of the log-normal distribution for $\alpha = 0$ and $\beta = 1$. Probabilities concerning random variables having a log-normal distribution can be calculated from the previously employed tables (see Chapter 9) of the normal distribution.

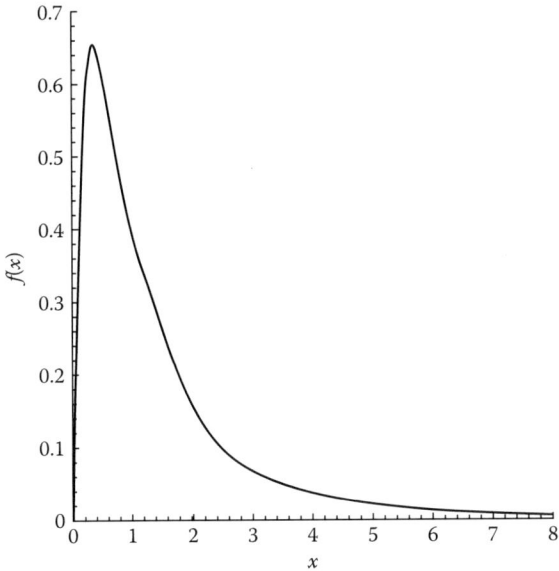

FIGURE 10.1 Log-normal pdf for $\alpha=0$, $\beta=1$.

If X has a log-normal distribution with parameters α and β, then $\ln X$ has a normal distribution with $\mu=\alpha$ and $\sigma=\beta$. Probabilities concerning X can therefore be converted into equivalent probabilities concerning $\ln X$.

For example, suppose that X has a log-normal distribution with $\alpha=2$ and $\beta=0.1$. Then,

$$P(6 < X < 8) = P(\ln 6 < \ln X < \ln 8)$$

$$= \left[\frac{\ln 6 - 2}{0.1} < \frac{\ln X - 2}{0.1} < \frac{\ln 8 - 2}{0.1} \right]$$

$$= P(-2.08 < Z < 0.79)$$

$$= (0.5 - 0.019) + (0.5 - 0.215)$$

$$= 0.481 + 0.285$$

$$= 0.78 = 78\%$$

Estimates of the parameters α and β in the pdf of a random variable X having a log-normal distribution can be obtained from a sample of observations on X by making use of the fact that $\ln X$ is normally distributed with mean α and standard deviation β. Therefore, the mean and standard deviation of the natural logarithms of the sample observations on X furnish estimates of α and β. To illustrate this procedure, suppose the time to failure T, in thousands of hours, was observed for a sample of five pumps at a water treatment plant. The observed values of T were 8, 11, 16, 22, and 34. Under the assumption that T has a log-normal distribution, the natural logs of the observation constitute a sample from a normal population. One first obtains the natural logs of the given observations on T, time to failure:

$\ln 8 = 2.08$ $\ln 16 = 2.77$ $\ln 34 = 3.53$ $\ln 11 = 2.40$ $\ln 22 = 3.09$

The mean, μ, and the standard deviation, σ, of these results can now be computed.

$$\mu = \sum_{i=1}^{5} \frac{\ln T_i}{5} = 2.77$$

$$\sigma = \sqrt{\sum_{i=1}^{5} \frac{\left[\ln T_i - 2.77\right]^2}{4}} = 0.57$$

The terms α and β in the pdf of T can be obtained directly from μ and σ, respectively.

Estimate of $\alpha = 2.77$

Estimate of $\beta = 0.57$

The aforementioned results can be employed to estimate that a motor lasts less than 5000 h. Once again, convert the required probability concerning T into a probability about $\ln T$:

$$P(T < 5) = P(\ln T < \ln 5) = P(\ln T < 1.61)$$

Treating $\ln T$ as a random variable that is normally distributed with mean α and standard deviation β, the required probability can be obtained using the standard normal table (see previous chapter).

$$P(\ln T < 1.61) = P\left(\left[\frac{\ln T - \alpha}{\beta}\right] < \left[\frac{1.61 - \alpha}{\beta}\right]\right)$$

$$= P\left(Z < \frac{1.61 - 2.77}{0.57}\right)$$

$$= P(Z < -2.04)$$

$$= 0.021 = 2.1\%$$

One can also calculate the probability that a pump lasts more than 5000 h. This result can be immediately deduced from

$$P(T > 5) = P(\ln T > 1.61) = 1 - P(T < 5)$$

$$= 1.0 - 0.021$$

$$= 0.979 = 97.9\%$$

Finally, one can proceed to calculate the probability that the pump lasts more than 10,000 h. The describing equation for this scenario becomes

$$P(T > 10) = P(\ln T > \ln 10) = P(\ln T < 2.303)$$

Converting to a standard normal variable,

$$P(\ln T > 2.303) = P\left(\left[\frac{\ln T - \alpha}{\beta}\right] > \left[\frac{2.303 - \alpha}{\beta}\right]\right)$$

$$= P\left(Z > \frac{2.303 - 2.77}{0.57}\right)$$

$$= P(Z > -2.434)$$

$$= 0.9926 = 99.26\%$$

The log-normal distribution has been employed to characterize occupational risk with time. It describes a distribution of data where there are many measurements with lower values and fewer measurements with high values. This distribution can be used to describe a relatively constant measurement, which is occasionally punctuated by higher values due to cyclic variations that arise in epidemiology studies, particularly for dose–response analysis. Since the log-normal distribution is also characterized by a geometric mean and a geometric standard deviation, the 95th percentile has been used as an indicator of peak values. This 95th percentile value is usually an important statistic for those chemicals that produce primarily acute toxicological effects.

The log-normal distribution, as noted, has been employed as an appropriate model in a wide variety of situations from environmental management to biology to economics. Additional applications include the distribution of personal incomes, inheritances, bank deposits, and also the distribution of organism growth subject to many small impurities. Rosenkrantz[3] has discussed the log-normal distribution in terms of a model for the distribution of stock prices. The unusual feature of this model is that it is the *logarithm* of the stock price—not the stock price itself—that is normally distributed. The author provides an extensive treatment of this topic plus illustrative examples. Perhaps the primary environmental application of the log-normal distribution has been to represent the distribution for particle sizes in gaseous emissions from many industrial processes.

ILLUSTRATIVE EXAMPLES

Illustrative Example 10.1

The failure rate per year, Y, of a coolant recycle pump in a wastewater treatment plant has a log-normal distribution. If ln Y has a mean of 2.0 and a variance of 1.5, find P(0.175 < Y < 1).

Solution

If Y has a log-normal distribution, $\ln Y$ has a normal distribution with a mean of 2.0 and a standard deviation $\sigma = 1.5^{1/2} = 1.22$. Therefore,

$$P(0.175 < Y < 1) = P(\ln 0.175 < \ln Y < \ln 1)$$

$$= P\left(\left[\frac{\ln 0.175 - 2}{1.22}\right] < \frac{\ln Y - 2}{1.22} < \left[\frac{\ln 1 - 2}{1.22}\right]\right)$$

$$= P(-3.07 < Z < -1.64)$$

$$= 0.1587 - 0.0011$$

$$= 0.1576 = 15.76$$

Illustrative Example 10.2

The data in Table 10.1 represent the time in microseconds between requests for a certain process service on a computer network at a landfill operation. Construct a frequency distribution for the data. Assuming that interrequest time has a log-normal distribution, obtain the theoretical frequencies for each class of the frequency distribution.

Solution

A frequency distribution is a table showing the number of observations in each of a succession of intervals called *classes*, as arbitrarily selected in Table 10.2. First compute the mean, μ, and standard deviation, σ, of the natural logs of the observations. Employ the results provided from Table 10.3. The mean, μ, is therefore

$$\mu = \sum_{i=1}^{50} \frac{\ln X_i}{50} = \frac{446.51}{50} = 8.93$$

TABLE 10.1

Computer Network Data

114,462	10,280	2,654	6,761	8,111
5,437	14,691	4,605	9,405	15,184
4,866	4,789	11,944	6,919	5,547
1,439	1,333	18,270	35,632	17,783
13,017	32,145	7,310	1,812	15,078
4,138	7,361	9,405	4,277	2,592
1,594	39,577	3,820	6,925	6,974
1,422	6,063	5,432	6,003	27,778
36,938	15,615	2,904	8,840	3,711
10,829	5,575	6,634	3,674	5,825

TABLE 10.2
Computer Network Frequency Distribution

Interrequest Time (µs)	Frequency
0–2,499	5
2,500–4,999	11
5,000–9,999	18
10,000–19,999	10
20,000–39,999	5
40,000–79,999	0
80,000–159,999	1
Total = 50	Total = 50

TABLE 10.3
Natural Log Frequency

X	ln X	X	ln X	X	ln X
114,462	11.65				
5,437	8.60	6,063	8.71	1,812	7.5
4,866	8.50	15,615	9.66	4,277	8.36
1,439	8.49	5,575	8.63	6,925	8.84
13,017	9.47	2,654	7.88	6,003	8.70
4,138	8.33	4,605	8.43	8,840	9.09
1,594	7.37	11,944	9.39	3,674	8.21
		18,270	9.81	8,111	9.00
1,422	7.26	7,310	8.90	15,184	9.63
36,938	10.52	7,405	9.15	5,547	8.62
10,829	9.29	3,820	8.25	17,783	9.79
10,280	9.24	5,432	8.60	15,078	9.62
14,691	9.59	2,904	7.97	2,592	7.86
4,789	8.47	6,634	8.80	6,974	8.85
1,333	7.20	6,761	8.82	27,778	10.23
32,145	10.38	9,405	9.15	3,711	8.22
7,361	8.90	6,919	8.84	5,825	8.67
39,577	10.59	35,632	10.48		

The standard deviation is

$$\sigma = \sqrt{\sum_{i=1}^{50} \frac{\left[\ln X_i - 8.93\right]^2}{49}} = 0.91$$

Estimate α and β from the mean and standard deviation values calculated above.

Estimate of $\alpha = 8.93$

Estimate of $\beta = 0.91$

Convert the probabilities associated with each class of the frequency distribution into corresponding probabilities about the natural logs of the endpoint of each class.

$$P(0 < X < 2,499) = P(-\infty < \ln X < \ln 2,499) = P(\infty < \ln X < 7.82)$$

$$P(2,500 < X < 4,999) = P(\ln 2,500 < \ln X < \ln 4,999) = P(7.82 < \ln X < 8.52)$$

$$P(5,000 < X < 9,999) = P(\ln 5,000 < \ln X < \ln 9,999) = P(8.52 < \ln X < 9.21)$$

$$P(10,000 < X < 19,999) = P(\ln 10,000 < \ln X < \ln 19,999) = P(9.21 < \ln X < 9.90)$$

$$P(20,000 < X < 39,999) = P(\ln 20,000 < \ln X < \ln 39,999) = P(9.90 < \ln X < 10.60)$$

$$P(40,000 < X < 79,999) = P(\ln 40,000 < \ln X < \ln 79,999) = P(10.60 < \ln X < 11.29)$$

$$P(80,000 < X < 159,999) = P(\ln 80,000 < \ln X < \ln 159,999) = P(11.29 < \ln X < 11.98)$$

Treating $\ln X$ as a random variable normally distributed with a mean of 8.93 and standard deviation of 0.9, evaluate the probabilities in the preceding data using the standard normal table.

$$P(-\infty < \ln X < 7.82) = P\left(-\infty < \left[\frac{\ln X - 8.93}{0.91}\right] < \left[\frac{7.82 - 8.93}{0.91}\right]\right)$$

$$= P(-\infty < Z < -1.22)$$

$$= 0.111$$

Similarly,

$$P(7.82 < \ln X < 8.52) = P(-1.22 < Z < -0.45)$$

$$= 0.326 - 0.111$$

$$= 0.215$$

$$P(8.52 < \ln X < 9.21) = P(-0.45 < Z < 0.31)$$

$$= 0.296$$

$$P(9.21 < \ln X < 9.90) = P(0.31 < Z < 1.07)$$

$$= 0.236$$

$$P(9.90 < \ln X < 10.60) = P(1.07 < Z < 1.84)$$

$$= 0.109$$

TABLE 10.4

Interrequest Frequency Results

Interrequest Time (µs)	Frequency	Probabilities	Theoretical Frequency
0–2,499	5	0.111	5.6
2,500–4,999	11	0.215	10.8
5,000–9,999	18	0.296	14.8
10,000–19,999	10	0.231	11.8
20,000–39,999	5	0.109	5.5
40,000–79,999	0	0.028	1.4
80,000–159,999	1	0.005	0.3

$$P(10.60 < \ln X < 11.29) = P(1.84 < Z < 2.59)$$

$$= 0.028$$

$$P(11.29 < \ln X < 11.98) = P(2.59 < Z < 3.35)$$

$$= 0.005$$

Obtain the theoretical frequency for each class by multiplying each of these probabilities by 50, the total tabulated frequency. Results are presented in Table 10.4.

Illustrative Example 10.3[4]

Particulate discharges from an operation, usually to the atmosphere, consist of a size distribution ranging anywhere from extremely small particles (less than 1 µm) to very large particles (greater than 100 µm). Particle size distributions are usually represented by a cumulative weight fraction curve in which the fraction of particles less than or greater than a certain size is plotted against the dimension of the particle.

To facilitate recognition of the size distribution, it is useful to plot a size–frequency curve. The size–frequency curve shows the number (or weight) of particles present for any specified diameter. Because most dusts are comprised of an infinite range of particle sizes, it is first necessary to classify particles according to some consistent pattern. The number or weight of particles may then be defined as that quantity within a specified size range having finite boundaries and typified by some average diameter.

The shape of the curves obtained to describe the particle size distribution generally follows a well-defined form. If the data include a wide range of sizes, it is often better to plot the frequency (i.e., number of particles of a specified size) against the logarithm of the size. In most cases, an asymmetrical or "skewed" distribution exists; normal probability equations do not apply to this distribution. Fortunately, in most instances, the symmetry can be restored if the logarithms of the sizes are substituted for sizes. The curve is then said to be logarithmic normal (or log-normal) in distribution. Plotting particle diameter versus cumulative percentage therefore generates cumulative distribution plots described in Chapter 4. For log-normal distributions, plots of particle diameter versus either percent less than stated size (% LTSS) or percent greater than stated size (% GTSS) produce straight lines on log-probability coordinates. The next three examples also address this application.

You have been requested to determine if a particle size distribution is log-normal. Data are provided in Table 10.5.

Solution

Cumulative distribution information can be obtained from the calculated results provided in Table 10.6.

The cumulative distribution can be plotted on log-probability graph paper. The cumulative distribution curve is shown in Figure 10.2. Because a straight line is obtained on log-normal coordinates, the particle size distribution is log-normal.

TABLE 10.5
Particle Size Data

Particle Size Range, d_p (μm)	Distribution (μg/m³)
<0.62	25.5
0.62–1.0	33.15
1.0–1.2	17.85
1.2–3.0	102.0
3.0–8.0	63.75
8.0–10.0	5.1
>10.0	7.65
Total	255.0

TABLE 10.6
Cumulative Distribution Data

Particle Size Range, d_p (μm)	Total (%)	Cumulative % GTSS[a]
<0.62	10	90
0.62–1.0	13	77
1.0–1.2	7	70
1.2–3.0	40	30
3.0–8.0	25	5
8.0–10.0	2	3
>10.0	3	0

[a] %GTSS represents the percent greater than stated size, where the stated size is the upper limit of the corresponding particle size range. Thus, 99.6% of the particles have a size equal to or greater than 2 μm.

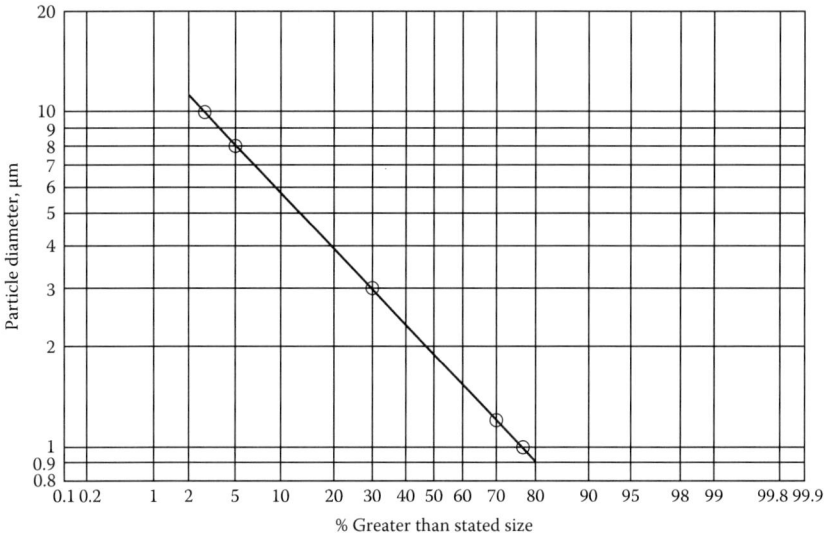

FIGURE 10.2   Cumulative size distribution for Application 10.3.

Illustrative Example 10.4

Refer to the previous example. Estimate the standard deviation from the size distribution information available. Use the 84.14% particle size value.

Solution

The use of probability plots is of value when the arithmetic or geometric mean is required because these values may be read directly from the 50% point on a logarithmic probability plot. By definition, the size corresponding to the 50% point on the probability scale is the *geometric mean diameter*. The geometric standard deviation is given (for % LTSS) by

$$\sigma = \frac{84.14\% \text{ size}}{50\% \text{ size}}$$

For % GTSS,

$$\sigma = \frac{50\% \text{ size}}{84.14\% \text{ size}}$$

The mean, as read from the 50% GTSS point on the graph in Figure 10.3, is approximately 1.9 μm. A value of 1.91 μm is obtained from an expanded plot. From the diagram, the particle size corresponding to the 84.14% point is

$$d_p\left(84.14\%\right) = 0.75 \ \mu m$$

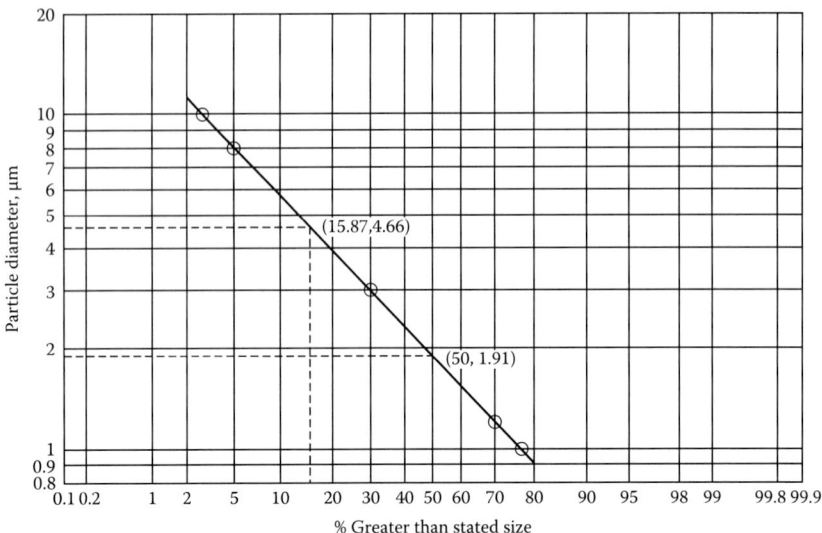

FIGURE 10.3   Cumulative distribution curve for Illustrative Example 10.4.

The standard deviation may now be calculated. Using the equation provided above,

$$\sigma = \frac{d_p(50\%)}{d_p(84.14\%)}$$
$$= \frac{1.91}{0.75}$$
$$= 2.59 \ \mu m$$

Illustrative Example 10.5

The following cumulative particle size data is provided in Table 10.7. Estimate the mean and standard deviation.

Solution

A plot (see Figure 10.4) on a log-probability paper yields a straight line. Therefore, the distribution is again log-normal. To determine the geometric mean diameter, one can read the 50% size:

$$d_p(50\%) = 10.5 \ \mu m$$

The standard deviation may now be calculated. By definition,

$$\sigma = \frac{d_p(50\%)}{d_p(84.13\%)}$$
$$= \frac{10.5}{5.5}$$
$$= 1.9 \ \mu m$$

TABLE 10.7
Particle Size Concentration Data

Particle Size Range (μm)	Concentration (μg/m³)	Weight in Size Range (%)	Cumulative % GTSS[a]
0–2	0.8	0.4	99.6
2–4	12.2	6.1	93.5
4–6	25.0	12.5	81.0
6–10	56	28	53.0
10–20	76	38	15.0
20–40	27	13.5	1.5
>40	3	1.5	—

[a] %GTSS represents the percent greater than stated size, where the stated size is the upper limit of the corresponding particle size range. Thus, 99.6% of the particles have a size equal to or greater than 2 μm.

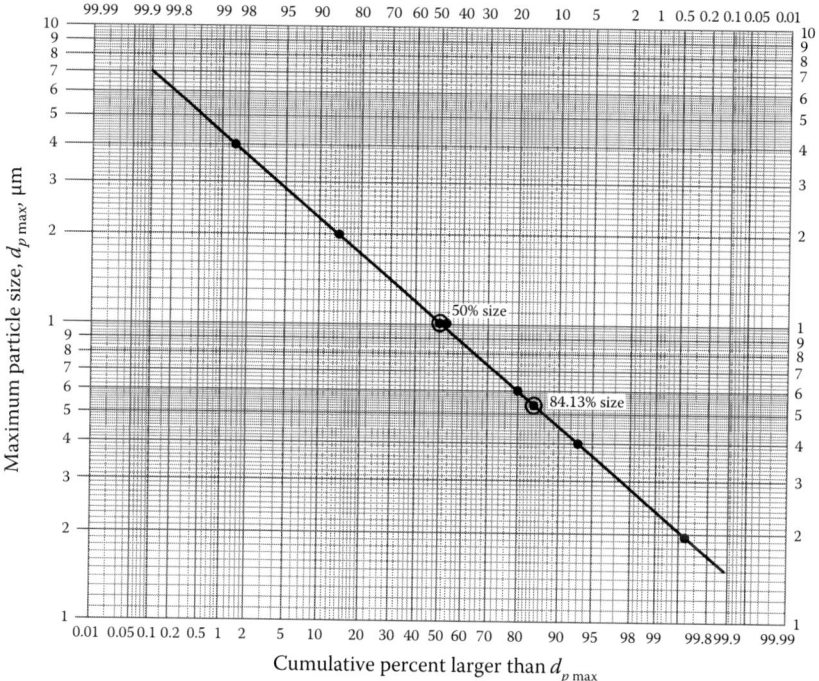

FIGURE 10.4   Log-probability distribution for data from Illustrative Example 10.5.

Illustrative Example 10.6

Given the Anderson 2000 sampler data from a coal-fired boiler, an environmental student has been requested to plot a cumulative distribution curve on log-probability paper and determine the mean particle diameter and geometric standard deviation of the fly ash discharge. Pertinent data are provided in Table 10.8. Note that

$$\text{Sampler volumetric flow rate}, q = 0.5 \text{ cfm}$$

See also Figure 10.5 for aerodynamic diameter[4] versus flow rate data from an Anderson sampler.

TABLE 10.8
Anderson 2000 Sampler Data

Plate Number	Tare Weight (g)	Final Weight (g)
0	20.48484	20.48628
1	21.38338	21.38394
2	21.92025	21.92066
3	21.55775	21.55817
4	11.40815	11.40854
5	11.61862	11.61961
6	11.76540	11.76664
7	20.99617	20.99737
Backup filter	0.20810	0.21156

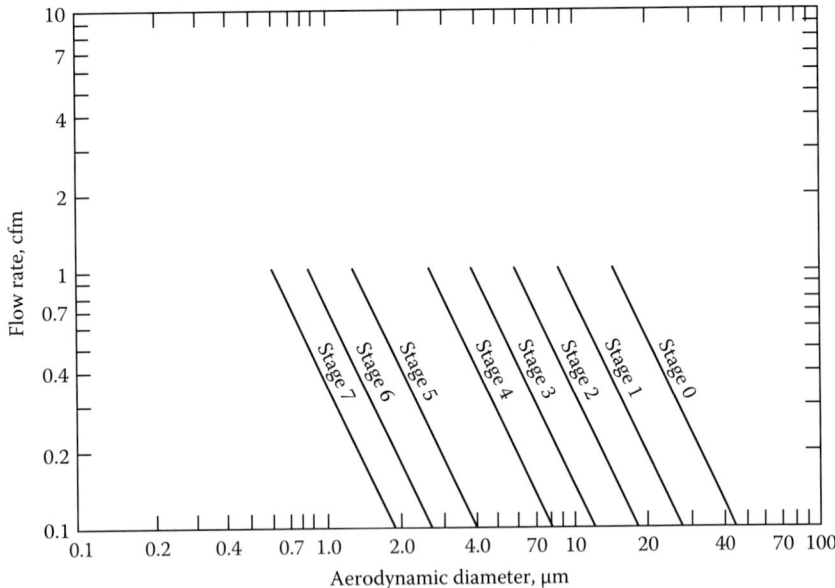

FIGURE 10.5 Aerodynamic diameter vs. flow rate through an Anderson sampler for an impaction efficiency of 95%.

TABLE 10.9
Anderson 2000 Sampler Calculations

Plate Number	Net Weight (g)	Percentage of Total Weight
0	1.44	14.2
1	0.56	5.5
2	0.41	4.1
3	0.42	4.2
4	0.39	3.9
5	0.99	9.8
6	1.24	12.3
7	1.20	11.9
Backup filter	3.46	34.2
Total	10.11	100.0

Solution

Table 10.9 provides the net weight and percentage of total weight. A sample calculation (for plate 0) follows:

$$Net\ weight = Final\ weight - Tare\ weight$$

$$= 20.48628 - 20.48484$$

$$= 1.44 \times 10^{-3}\ g = 1.44\ mg$$

$$Percentage\ of\ total\ weight = (net\ weight\ /\ total\ net\ weight)\,(100)$$

$$= (1.44/10.11)\,(100\%)$$

$$= 14.2\%$$

Calculate the cumulative percentage for each plate. Again for plate 0,

$$Cumulative\ \% = 100 - 14.2$$

$$= 85.8\%$$

For plate 1,

$$Cumulative\ \% = 100 - (14.2 + 5.5)$$

$$= 80.3\%$$

Table 10.10 shows the cumulative percentage for each plate.
    Using the Anderson graph shown in Figure 10.5, determine the 95% aerodynamic diameter at $q = 0.5$ cfm for each plate (stage). Table 10.11 shows the 95%

TABLE 10.10

Anderson Sampler Cumulative Percentage Data

Plate No.	Cumulative Percentage
0	85.8
1	80.3
2	76.2
3	72.0
4	68.1
5	58.3
6	46.0
7	34.1
Backup filter	—

TABLE 10.11

Anderson Sampler Aerodynamic Data

Plate No.	95% Aerodynamic Diameter (µm)
0	20.0
1	13.0
2	8.5
3	5.7
4	3.7
5	1.8
6	1.2
7	0.78

aerodynamic diameter for each plate. The cumulative distribution curve is provided on the log-probability coordinates in Figure 10.6.

The mean particle diameter is the particle diameter, Y, corresponding to a cumulative percentage of 50.

$$\text{Mean particle diameter} = Y_{50} = 1.6 \ \mu m$$

The distribution appears to approach log-normal behavior. The particle diameter at a cumulative percentage of 84.13 is

$$Y_{84.13} = 15 + \mu m \approx 15 \ \mu m$$

Therefore, the geometric standard deviation is approximately

$$\sigma_G = Y_{84.13} / Y_{50}$$

$$= 15 / 1.6$$

$$= 9.4 \ \mu m$$

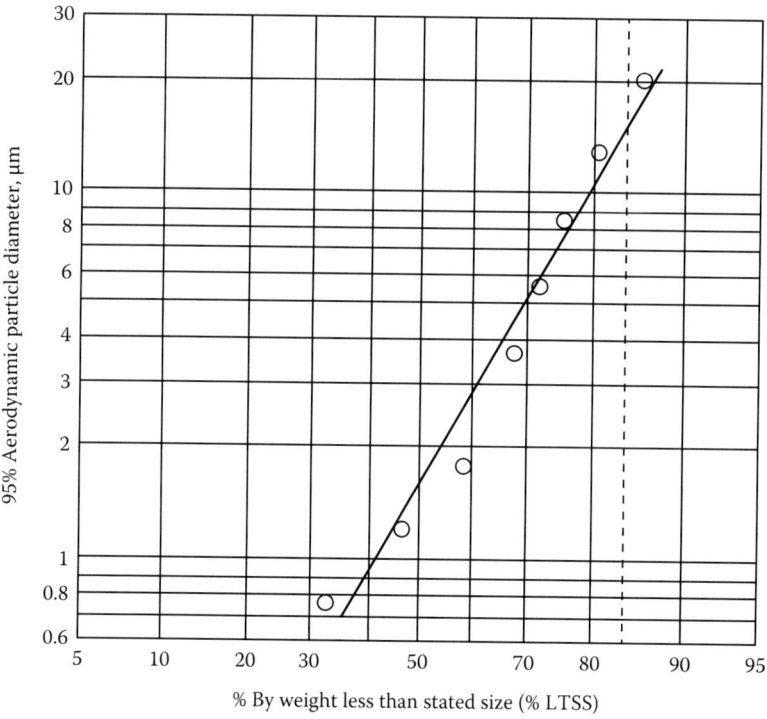

% By weight less than stated size (% LTSS)

FIGURE 10.6 % LTSS for Illustrative Example 10.7.

Illustrative Example 10.7

Normalized biological oxygen demand (BOD) levels in an estuary during the past 10 years are summarized in Table 10.12. If the BOD levels are assumed to follow a log-normal distribution, predict the level that would be exceeded only once in 100 years.

Solution

For this case, refer to Table 10.13. Based on the calculations presented in Table 10.13

$$\bar{X} = \frac{\sum X}{n} = \frac{38.86}{10} = 3.886$$

$$s^2 = \frac{\sum X^2 - \left(\sum X\right)^{2/n}}{n-1}$$

$$= \frac{156.78 - \left(38.86\right)^{2/10}}{10-1}$$

$$= 0.64$$

TABLE 10.12

Estuary BOD Data

Year	1	2	3	4	5	6	7	8	9	10
BOD Level	23	38	17	210	62	142	43	29	71	31

Note: BOD, biological oxygen demand.[5]

TABLE 10.13

BOD Calculations

Year (Y)	BOD Level	$X = \ln$ BOD	X^2
1	23	3.13	9.83
2	38	3.64	13.25
3	17	2.83	8.01
4	210	5.35	28.62
5	62	4.13	17.06
6	142	4.96	24.60
7	43	3.76	14.14
8	29	3.37	11.36
9	71	4.26	18.15
10	31	3.43	11.76
Total	—	38.86	156.78

and

$$s = 0.80$$

For this test, with $Z = 2.327$ for the 99% value,

$$Z = \frac{X - \bar{X}}{s}$$

$$2.327 = \frac{X - 3.886}{0.80}$$

Solving for X yields

$$X = 5.748$$

For this log-normal distribution,

$$X = \ln(\text{BOD})$$

$$X = 5.748 = \ln(\text{BOD})$$

$$\text{BOD} = 313$$

REFERENCES

1. L. Theodore and F. Taylor, *Probability and Statistics*, Theodore Tutorials (originally published by USEPA, RTP, NC), East Williston, NY, 1996.
2. S. Shaefer and L. Theodore, *Probability and Statistics Applications in Environmental Science*, CRC Press/Taylor & Francis Group, Boca Raton, FL, 2007.
3. W. Rosenkrantz, *Probability and Statistics for Science, Engineering, and Finance*, CRC Press/Taylor & Francis Group, Boca Raton, FL, 2009.
4. L. Theodore, *Air Pollution Control Equipment Calculations*, John Wiley & Sons, Hoboken, NJ, 2008.
5. M.K. Theodore and L. Theodore, *Introduction to Environmental Management*, CRC Press/Taylor & Francis Group, Boca Raton, FL, 2010.

11 Exponential Distribution

INTRODUCTION[1,2]

This chapter and the next introduce the two probability distributions—exponential and Weibull—that have received the most attention in environmental risk analysis/ assessment studies. In addition to these two distributions, the chapters introduce the general subjects of series and parallel systems plus reliability relations (briefly discussed in Chapter 9). These two topics are introduced in this chapter with illustrative examples complementary to the presentation.

Exponents, exponential functions, exponential derivatives, and exponential integrals find application in engineering and science. But their use in environmental risk calculations finds even wider applications. (This last statement particularly applies to the Weibull distribution, a topic that will receive treatment in the next chapter.) It is for this reason that the next paragraph precedes the material on the exponential distribution.

Several key exponential relationships are provided in Table 11.1. Some key exponential integral forms are provided in Table 11.2. Some key exponential integrals in closed form follow in Table 11.3.

Onto the topic of concern in this chapter. Exponential distribution is an important distribution in that it represents the distribution of the time (usually) required for a single event from a Poisson process to occur. In particular, in sampling from a Poisson distribution with parameter μ, the probability that no event occurs during $(0, t)$ is $e^{-\lambda t}$. Consequently, the probability that an event will occur during $(0, t)$ is

$$F(t) = 1 - e^{-\lambda t} \tag{11.1}$$

This represents the cumulative distribution function (cdf) of t. One can therefore show that the probability distribution function (pdf) is

$$f(t) = e^{-\lambda t} \tag{11.2}$$

Note that the parameter $1/\lambda$ (sometimes denoted as μ) is the expected value (see Chapter 4). Normally, the reciprocal of this value is specified and represents the expected value of $f(t)$.

Because the exponential function appears in the expression for both the pdf and cdf, the distribution is justifiably called the *exponential distribution*. A typical pdf of x plot is provided in Figure 11.1. Alternatively, the cumulative exponential distribution can be obtained from the pdf (with x replacing t):

$$F(x) = \int_0^x \lambda e^{-\lambda x} dx = 1 - e^{-\lambda x} \tag{11.3}$$

TABLE 11.1

Exponential Relationships

$a^{-n} = 1/a^n \quad a \neq 0$

$\sqrt{ab} = \sqrt{a}\sqrt{b}$

$(ab)^n = a^n b^n$

$(a^n)^m = a^{nm}$

$a^{n+m} = a^n a^m$

$(a^n)^{1/m} = a^{n-m}$

$\sqrt[n]{a} = a^{1/n} \quad \text{if } a > 0$

$a^{m/n} = (a^m)^{1/n} = \sqrt[n]{a^m}, \quad a > 0$

$a^0 = 1 \, (a \neq 0)$

$0^a = 0 \, (a \neq 0)$

$(ab)^x = a^x b^x$

$b^x b^y = b^{x+y}$

$(b^x)^y = b^{xy}$

$\log ab = \log a + \log b, \quad a > 0, \, b > 0$

$\log a^n = n \log a$

$\log(a/b) = \log a - \log b$

$\log \sqrt[n]{a} = 1/n \log a$

One often encounters a random variable's conditional failure density or hazard function, $g(x)$, in statistical, reliability, and risk applications. In particular, $g(x) \, dx$ is the probability that a "product" will fail during $(x, x+dx)$ under the condition that it had not failed before time x. Consequently,

$$g(x) = \frac{f(x)}{1 - F(x)} \tag{11.4}$$

If the probability density function $f(x)$ is exponential, with parameter λ, it follows from Equations 11.2 and 11.3 that

$$g(x) = \frac{\lambda e^{-\lambda x}}{1 - (1 - e^{-\lambda x})}$$

$$= \frac{\lambda e^{-\lambda x}}{e^{-\lambda x}}$$

$$g(x) = \lambda \tag{11.5}$$

TABLE 11.2

Exponential Integral Forms

$$\int e^x dx = e^x$$

$$\int e^{-x} dx = -e^{-x}$$

$$\int e^{ax} dx = \frac{e^{ax}}{a}$$

$$\int x e^x dx = \frac{e^{ax}}{a^2}(ax-1)$$

$$\int \frac{e^{ax} dx}{x} = \log x + \frac{ax}{1!} + \frac{a^2 x^2}{2 \cdot 2!} + \frac{a^3 x^3}{3 \cdot 3!} + \cdots$$

$$\int \frac{e^{ax}}{x^m} dx = \frac{1}{m-1} \frac{e^{ax}}{x^{m-1}} + \frac{a}{m-1} \int \frac{e^{ax}}{x^{m-1}} dx$$

$$\int e^{ax} \log x \, dx = \frac{e^{ax} \log x}{a} - \frac{1}{a} \int \frac{e^{ax}}{x} dx$$

$$\int \frac{dx}{1+e^x} = x - \log(1+e^x) = \log \frac{e^x}{1+e^x}$$

$$\int \frac{dx}{a+be^{px}} = \frac{x}{a} - \frac{1}{ap} \log(a+be^{px})$$

Equation 11.5 indicates that the failure probability is constant, irrespective of time. It implies that the probability that a component whose time-to-failure distribution is exponential fails in an instant during the first hour of its life is the same as its failure probability during an instant in the thousandth hour—presuming it has survived up to that instant. It is for this reason that the parameter λ is usually referred to in life-test applications as the *failure rate*. This definition generally has meaning *only* with an exponential distribution. Other failure rate definitions and applications will be discussed later in the next chapter.

The aforementioned natural association with life-testing and the fact that it is very tractable mathematically makes the exponential distribution attractive as representing the life distribution of a complex system or several complex systems. It is the author's opinion that the exponential distribution is as prominent in reliability/risk assessment calculations and analysis as the normal distribution is in other branches of engineering and science. However, its role is probably not as significant as with the Weibull distribution (see Chapter 12).

It has been shown theoretically that this distribution provides a reasonable model for systems designed with a limited degree of redundancy and made up of many components, none of which have a high probability of failure. This is especially true when low component failure rates are maintained by periodic inspection and replacement or in situations in which failure is a function of outside phenomena

---

TABLE 11.3

Exponential Integrals—Closed Form

$$\int_0^\infty e^{-ax}dx = \frac{1}{a}; \quad (a > 0)$$

$$\int_0^\infty \frac{e^{-ax} - e^{-bx}}{x}dx = \log\frac{b}{a}; \quad (a,b > 0)$$

$$\int_0^\infty e^{-a^2x^2}dx = \frac{1}{2a}\sqrt{\pi} = \frac{1}{2a}\Gamma\left(\frac{1}{2}\right), \quad (a > 0)$$

$$\int_0^\infty xe^{-x^2}dx = \frac{1}{2}$$

$$\int_0^\infty x^2 e^{-x^2}dx = \frac{\sqrt{\pi}}{4}$$

$$\int_0^\infty x^{2n} e^{-ax^2}dx = \frac{1\cdot3\cdot5...(2n-1)}{2^{n+1}a^n}\sqrt{\frac{\pi}{a}}$$

$$\int_0^\infty x^m e^{-ax}dx = \frac{m!}{a^{m+1}}\left[1 - e^{-a}\sum_{r=0}^m \frac{a^r}{r!}\right]$$

$$\int_0^\infty e^{\left(-x^2 - \frac{a^2}{x^2}\right)}dx = \frac{e^{-2a}\sqrt{\pi}}{2}, \quad (a \geq 0)$$

$$\int_0^\infty e^{-nx}\sqrt{x}dx = \frac{1}{2n}\sqrt{\frac{\pi}{n}}$$

$$\int_0^\infty \frac{e^{-nx}}{\sqrt{x}}dx = \sqrt{\frac{\pi}{n}}$$

Note: The symbol Γ denotes the gamma function.

---

rather than a function of previous conditions. On the other hand, the exponential distribution often cannot represent individual component life (because of "infant mortalities" and wear-out patterns—see also Chapter 12), and it is sometimes questionable even as a system model.

Consider the following example. A young environmental engineer is interested in estimating the probability that a pump will survive at least three times its expected life. Assume the exponential distribution applies. Since the exponential distribution applies,

$$P(T) = e^{-\lambda t} \tag{11.6}$$

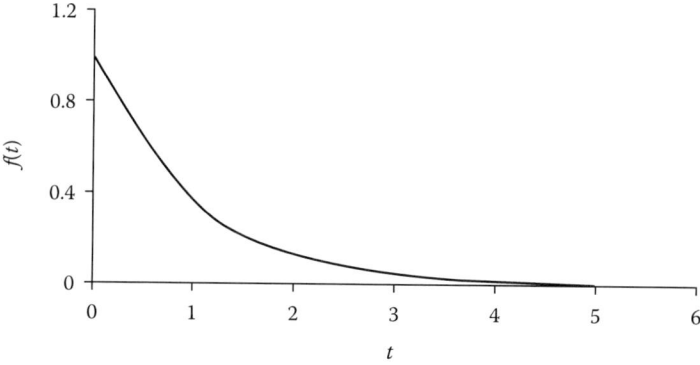

FIGURE 11.1 Exponential distribution.

with $\lambda = 1/a$ and $t = 3a$, where $a =$ expected life of the pump. Thus,

$$P(T > 3a) = e^{-\left(\frac{1}{a}\right)(3a)}$$
$$= e^{-3}$$
$$= 0.0498 = 4.98\%$$

Therefore, there is a 5% chance that the pump will survive past three times its expected life.

One can also calculate the probability that a pump will survive:

1. At least 5 times its expected life
2. At least 10 times its expected life

The describing equation

$$P(T) = e^{-\lambda t} \tag{11.6}$$

remains the same.

For case 1,

$$t = 5a$$

so that

$$P(T > 5a) = e^{-(1/a)(5a)}$$
$$= e^{-5}$$
$$= 0.0067 = 0.67\%$$

Similarly for case 2,

$$P(T > 10a) = e^{-(1/a)(10a)}$$

$$= e^{-10}$$

$$= 4.54 \times 10^{-5} = 4.54 \times 10^{-3}\%$$

As expected, the probability decreases with increasing survival time.

It should be noted that many systems, including those described by exponential and Weibull distributions, consisting of several components, can be classified as *series, parallel,* or a combination of both. However, the majority of industrial and process plants (units and systems) have a series of parallel configurations. A *series system* is one in which the entire system fails to operate if any one of its components fails to operate. If such a system consists of n components that function independently, then the reliability of the system is the product of the reliabilities of the individual components. (A detailed mathematical treatment of reliability relations is provided later in this chapter.) If R_s denotes the reliability of a series system and R_i denotes the reliability of the ith component where $i = 1, \ldots, n$, then

$$R_s = R_1 R_2 \cdots R_n$$

$$= \prod_{i=1}^{n} R_i \qquad (11.7)$$

A *parallel system* is one that fails to operate only if all its components fail to operate. If R_i is the reliability of the ith component, then $(1 - R_i)$ is the probability that the ith component fails; $i = 1, \ldots, n$. Assuming that all n components function independently, the probability that all n components fail is $(1 - R_1)(1 - R_2) \ldots (1 - R_n)$. Subtracting this product from unity yields the following formula for R_p, the reliability of a parallel system.

$$R_p = 1 - (1 - R_1)(1 - R_2)\ldots(1 - R_n)$$

$$= 1 - \prod_{i=1}^{n}(1 - R_i) \qquad (11.8)$$

The reliability formulas for series and parallel systems can be used to obtain the reliability of a system that combines features of a series and a parallel system as shown in Figure 11.2. Consider the system diagrammed in Figure 11.3. Components A, B, C, and D have for their respective reliabilities 0.90, 0.90, 0.80, and 0.90. The system fails to operate if A fails, if B and C both fail, or if D fails. One can now determine the reliability of the system.

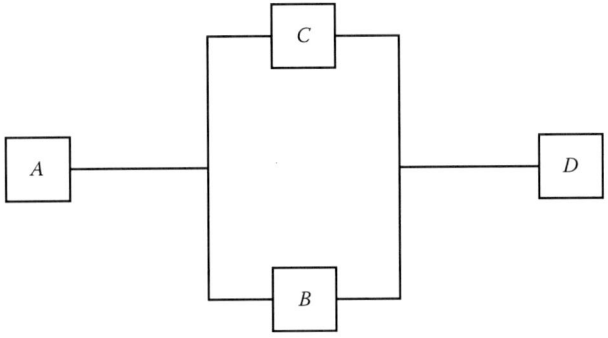

FIGURE 11.2   System with parallel and series components.

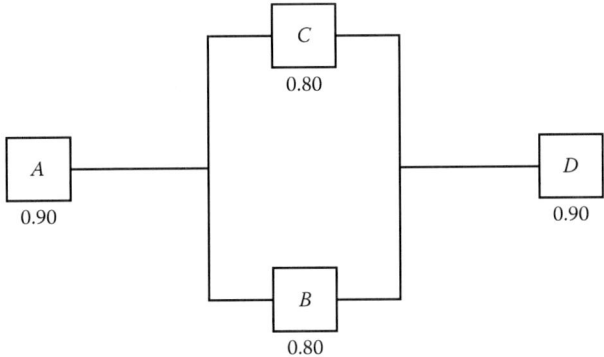

FIGURE 11.3   System with parallel and series component values.

Components B and C constitute a parallel subsystem connected in series to components A and D. The reliability of the parallel subsystem is obtained by applying Equation 11.8, which yields

$$R_p = 1 - (1 - 0.80)(1 - 0.80) = 0.96$$

The reliability of the system is then obtained by applying Equation 11.7, which yields

$$R_s = (0.90)(0.96)(0.90)$$

$$= 0.78 = 78\%$$

Consider also the reliability of the electrical system shown in Figure 11.4 using the reliabilities indicated under the various components. First identify the components connected in parallel. A and B are connected in parallel. D, E, and F are also connected in parallel. Then compute the reliability of each subsystem of the

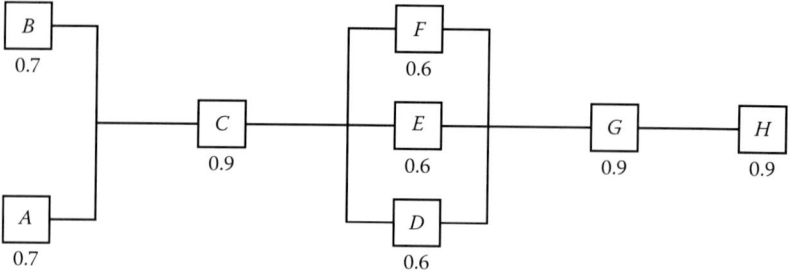

FIGURE 11.4 Diagram of an electrical system.

components connected in parallel. The reliability of the parallel subsystem consisting of components A and B is

$$R_p = 1 - (1 - 0.7)(1 - 0.7)$$

$$= 0.91 = 91\%$$

The reliability of the parallel subsystem consisting of components D, E, and F is

$$R_p = 1 - (1 - 0.6)(1 - 0.6)(1 - 0.6)$$

$$= 0.936 = 93.6\%$$

One may now multiply the product of the reliabilities of the parallel subsystems by the product of the reliabilities of the components to which the parallel subsystems are connected in series:

$$R_s = (0.91)(0.9)(0.936)(0.9)(0.9)$$

$$= 0.621 = 62.1\%$$

The reliability of the whole system is therefore 0.621 or 62.1%.

Finally, consider the following failure analysis. A military overseas flight is regarded as a series system with the following components: ground crew (A), cockpit crew (B), aircraft (C), weather conditions (D), and landing accommodations (E). The cockpit crew is viewed as a parallel system with the following components: captain (B_1), copilot (B_2), and flight engineer (B_3). Landing accommodations are viewed as a parallel system with the following components: scheduled airport (E_1), alternate landing sites (E_2 and E_3). Failure probabilities for the various components have been estimated as follows:

$$A = 0.001 \quad B_3 = 0.100 \quad E_1 = 0.001$$

$$B_1 = 0.001 \quad C = 0.001 \quad E_2 = 0.050$$

$$B_2 = 0.010 \quad D = 0.0001 \quad E_3 = 0.100$$

What is the probability of a successful flight?

First identify the components connected in parallel. B_1, B_2, and B_3 are connected in parallel. E_1, E_2, and E_3 are also connected in parallel. Compute the reliability of each subsystem of the components connected in parallel. The reliability of a parallel subsystem consisting of components B_1, B_2, and B_3 is

$$R_p = 1-(1-0.999)(1-0.99)(1-0.90) = 0.999999$$

The reliability of the parallel subsystem consisting of the components E_1, E_2, and E_3 is

$$R_p = 1-(1-0.999)(1-0.95)(1-0.90) = 0.999995$$

One may now multiply the product of the reliabilities of the parallel subsystems by the product of the reliabilities of the components to which the parallel subsystems are connected in series:

$$R_s = (0.999999)(0.999995)(0.999)(0.999)(0.9999) = 0.9979$$

The probability of a successful flight is therefore 0.9979 or 99.79%.

Onto reliability relations. One of the major applications in risk analysis involves reliability calculations. The reliability of a component will frequently depend on the length of time it has been in service. Let T, the time to failure, be the random variable having its pdf specified by $f(t)$. Then the probability that failure occurs in the time interval $(0, t)$ if given by[1]

$$F(t) = \int_0^t f(t)\, dt \tag{11.9}$$

Let the reliability of the component be denoted by $R(t)$, the probability that the component survives to time t. Therefore,

$$R(t) = 1 - F(t) \tag{11.10}$$

Equation 11.10 establishes the relationship between the reliability of a component and the cdf of its time to failure. The probability that a component will fail in the time interval $(t, t+\Delta t)$ is given by

$$P(t < T < t+\Delta t) = F(t+\Delta t) - F(t) \tag{11.11}$$

The conditional probability (see Chapter 3) that a component will fail in the time interval $(t, t+\Delta t)$, given that it has survived to time t, is

$$P\left[(t < T < \frac{t+\Delta t}{T > t}\right] = \frac{F(t+\Delta t) - F(t)}{P(T > t)} \tag{11.12}$$

Equation 11.12 is obtained by application of the definition of conditional probability. Noting that

$$R(t) = P(T > t) \tag{11.13}$$

and substituting in Equation 11.13 leads to

$$P\left[t < T < \frac{\Delta t}{T > t}\right] = \frac{F(t + \Delta t) - F(t)}{R(t)} \tag{11.14}$$

Division of both sides of Equation 11.14 by Δt yields

$$\frac{P[t < T < (t + \Delta t) / (T > t)]}{\Delta t} = \left[\frac{F(t + \Delta t) - F(t)}{\Delta t}\right]\left[\frac{1}{R(t)}\right] \tag{11.15}$$

Recall that $F'(t)$, the derivative of $F(t)$, is defined by

$$\lim_{\Delta t \to 0}\left[\frac{F(t + \Delta t) - F(t)}{\Delta t}\right] = F'(t) \tag{11.16}$$

By taking the limit of both sides of Equation 11.16 as Δt approached 0,

$$Z(t) = \frac{F'(t)}{R(t)} \tag{11.17}$$

where $Z(t)$ is defined by

$$Z(t) = \lim_{\Delta t \to 0} \frac{P[(t < T < t + \Delta t) / (T > t)]}{\Delta t} \tag{11.18}$$

The term $Z(t)$ is defined as the *failure rate* (also known as the *hazard rate*) of the component. Equation 11.18 establishes the relationship between failure rate reliability and the cdf of time to failure.

Using Equations 11.10 and 11.13, one can obtain an expression for the reliability in terms of the failure rate. Differentiating both sides of Equation 11.10 with respect to t yields

$$R'(t) = 0 - F'(t)$$
$$= -F'(t) \tag{11.19}$$

Substitution in Equation 11.17 yields

$$Z(t) = -\frac{R'(t)}{R(t)} \qquad (11.20)$$

Integrating both sides of Equation 11.20 between 0 and t yields

$$\int_0^t Z(t)\,dt = -[\ln R(t) - \ln R(0)] \qquad (11.21)$$

Since $R(t) = P(T > t)$, and $R(0) = 1$, Equation 11.21 becomes

$$\int_0^t Z(t)\,dt = -\ln R(t) \qquad (11.22)$$

Solving Equation 11.22 for $R(t)$ yields

$$R(t) = \exp\left[-\int_0^t Z(t)\,dt\right] \qquad (11.23)$$

the desired expression for the reliability in terms of failure rate.

The pdf of time to failure can also be expressed in terms of failure rate. Differentiating Equation 11.23 with respect to t yields

$$R'(t) = -Z(t)\exp\left[-\int_0^t Z(t)\,dt\right] \qquad (11.24)$$

Equation 11.10 may also be written (by differentiating both sides)

$$R'(t) = -f(t) \qquad (11.25)$$

Equation 11.24 can therefore be written as

$$f(t) = Z(t)\exp\left[-\int_0^t Z(t)\,dt\right] \qquad (11.26)$$

which represents the desired expression for the pdf in terms of failure rate.

Examples illustrating the application of these equations to real systems are provided in this and the next chapter. This chapter is limited to exponential distributions. However, in addition to the exponential, as well as the Weibull and normal distributions, several other probability distributions figure prominently in reliability calculations. These will receive treatment in Chapter 13.

Consider the following sample examples. A battery employed at an incineration site is deemed reliable if it operates for more than 500 h. The lives of the previous 11 batteries, in hours, were

$$501,\ 591,\ 621,\ 386,\ 942,\ 503,\ 201,\ 1013,\ 902,\ 32,\ 899$$

Assuming all batteries came from the same population, one notes that 3 of the 11 did not function beyond 500 h. Therefore, the reliability of the battery is simply given by

$$R = \frac{8}{11}$$

$$= 0.727 = 72.7\%$$

If two of these batteries are required in series for a retrofitted unit, one can estimate the reliability of the two batteries. Because the two batteries, with $R=0.727$, are connected in series,

$$R_s = R_1 R_2 \ldots R_n$$

$$= (0.727)\,(0.727)$$

$$= 0.529 = 52.9\% \tag{11.7}$$

If the two batteries are in a parallel system, applying Equation 11.8

$$R_p = 1 - (1 - R_1)(1 - R_2)$$
$$= 1 - (1 - 0.727)\,(1 - 0.727)$$
$$= 0.924 = 92.4\%$$

Employing the aforementioned information provided, and assuming battery life can be reasonably described by an exponential equation, one can estimate the reliability that a battery would survive 100 h. For this calculation, the average time to failure, t_f, is

$$t_f = \frac{(501 + 591 + 621 + 386 + 942 + 503 + 201 + 1013 + 902 + 32 + 899)}{11}$$

$$= \frac{6591}{11}$$

$$= 599\ \text{h}$$

For an exponential model,

$$f(t) = e^{-\lambda t} \tag{11.2}$$

where λ is the reciprocal of the average time to failure. Thus,

$$\lambda = \frac{1}{t_f} = \frac{1}{599} = 0.00167$$

The reliability for 100 h is therefore (from Equation 11.10)

$$R = e^{-(0.00167)(100)}$$

$$= e^{-0.167}$$

$$= 0.846 = 84.6\%$$

ILLUSTRATIVE EXAMPLES

Illustrative Example 11.1

The time to failure for a battery is presumed to follow an exponential distribution with $\lambda = 0.1$ (per year). What is the probability of a failure within the first year?

Solution
Refer to Equation 11.3 in the introduction to this chapter.

$$F(x) = \int_0^x \lambda e^{-\lambda x}dx = 1 - e^{-\lambda x}$$

For this case,

$$P(X \le 1) = \int_0^1 (0.1)e^{-(0.1)x}dx$$

$$= -\frac{0.1}{0.1}e^{-(0.1)x}\Big|_0^1$$

$$= -e^{-0.1} + e^0$$

$$= 1 - e^{-0.1}$$

$$= 0.095 = 9.5\%$$

Therefore, there is nearly a 10% probability that the battery will fail within the first year.

Illustrative Example 11.2

An electronic system consists of three components (1, 2, 3) connected in parallel. If the time to failure for each component is exponentially distributed and mean times of failures for components 1, 2, and 3, are 200, 300, and 600 days, respectively, determine the system reliability for 265 days.

Solution

For this series system, the probability of failure is (see also Equation 11.8)

$$P(F) = P(\text{all components fail})$$
$$= (1 - P_1)(1 - P_2)(1 - P_3)$$

where P_i is the probability of surviving 365 days. T the system reliability is

$$R = 1 - P(F)$$
$$= 1 - (1 - P_1)(1 - P_2)(1 - P_3)$$

Based on the data provided,

$$P_1 = e^{-\left(\frac{365}{200}\right)} = 0.161 = 16.1\%$$

$$P_2 = e^{-\left(\frac{365}{300}\right)} = 0.296 = 29.6\%$$

$$P_3 = e^{-\left(\frac{365}{600}\right)} = 0.544 = 54.4\%$$

Therefore, the system reliability is

$$R = 1 - (1 - 0.161)(1 - 0.296)(1 - 0.544)$$
$$= 0.731 = 73.1\%$$

Illustrative Example 11.3

A pumping system consists of four components. Three are connected in parallel, which in turn are connected downstream (in series) with another component. The arrangement is schematically shown in Figure 11.5. If the pumps have the same exponential failure rate, λ, of 0.5 (year)$^{-1}$, estimate the probability that the system will not survive for more than 1 year.

Solution

Based on the information provided, the pumping system fails when the three parallel components fail or when the downstream component fails. This is a combination of a parallel and series system. From Equation 11.10, the reliability is

$$R(t) = 1 - F(t) \tag{11.10}$$

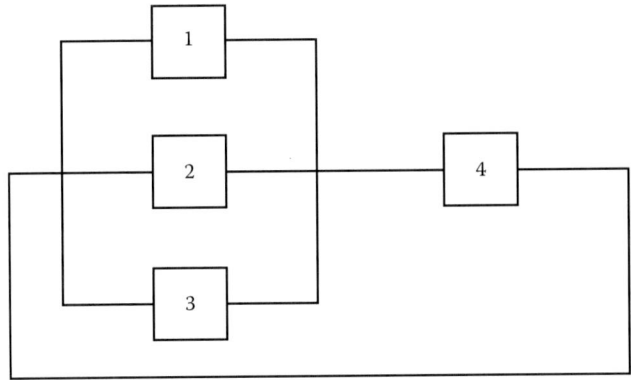

FIGURE 11.5 Pumping system (see Illustrative Example 11.3).

where $F(t)$ is the probability of failure between 0 and t. The reliability of the parallel system is[1]

$$R_p = 1 - (1 - R_1)(1 - R_2)(1 - R_3); \quad R_i = R$$
$$= 1 - (1 - R)^3$$

For a series system[1] (see also Equation 11.7)

$$R_s = R_p R_4 = R_p R$$

where R_s also represents the overall system reliability.
Applying the exponential model gives

$$R(t) = 1 - F(t)$$
$$= 1 - (1 - e^{-\lambda t})$$
$$= e^{-\lambda t}$$
$$= R_1 = R_2 = R_3 = R_4 = R = e^{-(0.5)(1)}$$
$$R = 0.6065 = 60.7\%$$

Thus,

$$R_p = 1 - (1 - 0.6065)^3$$
$$= 1 - 0.0609$$
$$= 0.9391 = 94.0\%$$

and

$$R_s = (0.9391)(0.6065)$$
$$= 0.57 = 57\%$$

Illustrative Example 11.4

The probability that a thermometer in a hazardous waste incinerator will not survive for more than 36 months is 0.925; how often should the thermometer be replaced? Assume the time to failure is exponentially distributed and that the replacement time should be based on the thermometer's expected life.

Solution

This requires the calculation of μ in the exponential model with units of (month)$^{-1}$. Once again,

$$F(t) = 1 - e^{-\lambda t} \tag{11.1}$$

Based on the information provided

$$P(T \leq 36) = 0.925$$

or

$$0.925 = 1 - e^{-(\lambda)(36)}$$

Solving for λ gives

$$\lambda = -\frac{1}{36}\ln(1 - 0.925)$$

$$= 0.07195$$

Because the expected time (or life), $E(T)$ is (see also Chapter 4)

$$E(T) = \frac{1}{\lambda}$$

$$= \frac{1}{0.07195}$$

$$= 13.9 \, months$$

The thermometer should be replaced every 14 months.

Illustrative Example 11.5

Refer to the previous example. Determine when the thermometer should be replaced if the probability (fractional basis) of thermometer survival for 36 months is 0.95.

Solution

The describing equation remains

$$F(t) = 1 - e^{-\lambda t}$$

For $F(t) = 0.95$,

$$0.95 = 1 - e^{\lambda t}$$

Solving for λ gives

$$\lambda = -\frac{1}{36}\ln(1 - 0.95)$$

$$= 0.0832 = 8.32\%$$

The expected life is

$$E(T) = \frac{1}{\lambda}$$

$$= \frac{1}{0.0832}$$

$$= 12 \text{ months}$$

Illustrative Example 11.6

Refer to the previous example. Calculate the survival time if the survival probability is 0.99.

Solution

Following the same procedure,

$$\lambda = -\frac{1}{36}\ln(1 - 0.99)$$

$$= 0.128 = 12.8\%$$

and

$$E(T) = \frac{1}{\lambda}$$

$$= \frac{1}{0.128}$$

$$= 7.82 \text{ months}$$

As expected, the survival time has increased.

Illustrative Example 11.7

Consider the system shown in Figure 11.6. Determine the reliability, R, if the operating time for each unit is 5000 h. Components A and B have exponential failure rates, λ, of 3×10^{-6} and 4×10^{-6} failures per hour, respectively, where $R_i = e^{-\lambda_i t}$; $t =$ time, h. The term λ may once again be viewed as the reciprocal of the average time to failure.

Solution

Because this is a series system (see also Equation 11.7)

$$R_s = R_A R_B$$

As indicated above, the failure rate is given by

$$R = e^{-\lambda t}; \quad t = \text{time, h}$$

so that

$$R_A = e^{(3 \times 10^{-6})(5000)} = e^{0.015} = 0.9851$$

and

$$R_B = e^{(4 \times 10^{-6})(5000)} = e^{0.02} = 0.9802$$

Therefore,

$$R_s = (0.9851)(0.9802)$$

$$= 0.9656 = 96.6\%$$

Illustrative Example 11.8

Refer to the previous example. Recalculate the reliability of the system if the order is reversed, that is, A follows B.

Solution

Because

$$R_s = R_A R_B = R_B R_A$$

the reliability remains the same.

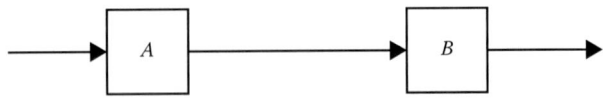

FIGURE 11.6 Exponential failure rate: Series system (see Illustrative Example 11.7).

Illustrative Example 11.9

Consider the system shown in Figure 11.7. Determine the reliability of this system employing the exponential failure rate information provided in the two previous examples.

Solution

Because this is a parallel system (see also Equation 11.8)

$$R_p = 1-(1-R_A)(1-R_B)$$

Employing the results from the previous example

$$R_p = 1-(1-0.9851)(1-0.9802)$$
$$= 1.0-(0.0149)(0.0198)$$
$$= 0.9997 = 99.97\%$$

Illustrative Example 11.10

Refer to the previous example. Recalculate the reliability if the two components are interchanged.

Solution

Because

$$R_p = 1-(1-R_A)(1-R_B)$$
$$= 1-(1-R_B)(1-R_A)$$

the reliability again remains the same.

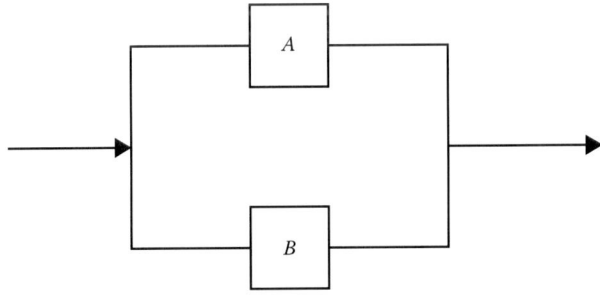

FIGURE 11.7   Exponential failure rate: Parallel system (see Illustrative Example 11.9).

Illustrative Example 11.11

Recalculate the battery reliability for 500 h. Employ the battery life data provided in the introduction to this chapter (just prior to the start of the illustrative examples) and assume the exponential distribution recorded applies, that is, $x = 0.00167(h)^{-1}$.

Solution

The same reliability equation is employed with $t = 500$ h. Therefore (as before),

$$R = e^{-(0.00167)(500)}$$

$$= 0.434 = 43.4\%$$

Illustrative Example 11.12

Explain the different results obtained in the previous example and the example provided in the Introduction for this chapter.

Solution

The reliability of the battery in the Introduction was 72.7%, whereas the reliability of the batter in the previous example was 43.4%. This difference arises because the distribution of failure times given in the Introduction did not come from an exponential distribution, that is, it did not come from a population that has a constant failure rate. An exponential distribution was assumed in the previous example.

REFERENCES

1. L. Theodore and F. Taylor, *Probability and Statistics*, Theodore Tutorials (originally published by USEPA, RTP, NC), East Williston, NY, 1996.
2. S. Shaefer and L. Theodore, *Probability and Statistics Applications in Environmental Science*, CRC Press/Taylor & Francis Group, Boca Raton, FL, 2007.

12 Weibull Distribution

INTRODUCTION[1,2]

Simply put, the Weibull distribution describes *failure rate* as a function of time. It has served the technical community for over 50 years; the chemical industry, refineries, the Pentagon, NASA, and so on, have been the beneficiaries. Thus, it is a key chapter in this book and is understandably one of the longest. The chapter not only reviews the traditional material in the literature on the Weibull distribution, but also includes illustrative examples concerned with (see also previous chapter) reliability relations plus series and parallel systems. Recent efforts of the author to improve on Weibull's work are also detailed.

Unlike the exponential distribution, the failure rate of equipment frequently exhibits three stages as: a break-in (BI) stage with a declining failure rate, a useful life stage characterized by a fairly constant failure rate, and a wear-out (WO) period characterized by an increasing failure rate. Many industrial parts and components follow this path. A failure rate curve exhibiting these three phases (see Figure 12.1) is called a *bathtub curve*.

In the case of the bathtub curve, failure rate during useful life is constant. Letting this constant be α and substituting it for $Z(t)$ in Equation 11.26 in the previous chapter yields

$$F(t) = \alpha \exp\left[-\int_0^t \alpha \, dt \right]$$

$$= \alpha \exp(-\alpha t); \quad t > 0$$

$$= \alpha e^{-\alpha t}$$

(12.1)

as the probability distribution function (pdf) of time to failure during the useful life stage of the bathtub curve. Equation 12.1 defines an exponential pdf (see Equation 11.2), that is, a special case of the pdf defining the Weibull distribution.

Weibull introduced the distribution, which bears his name principally on empirical grounds to represent certain life-test data. The Weibull distribution provides a mathematical model of all three stages of the bathtub curve. This is now discussed. An assumption about the failure rate that reflects all three stages of the bathtub stage is

$$Z(t) = \alpha \beta t^{\beta - 1}; \quad t > 0$$

(12.2)

where α and β are constants, referred to by some as shape parameters or curve-fitting parameters. For $\beta < 1$, the failure rate $Z(t)$ decreases with time. For $\beta = 1$, the failure rate is constant and equal to α. For $\beta > 1$, the failure rate increases with time. Using Equation 11.26 again to translate the assumption about failure rate into a corresponding assumption about the pdf of T, time to failure, one obtains

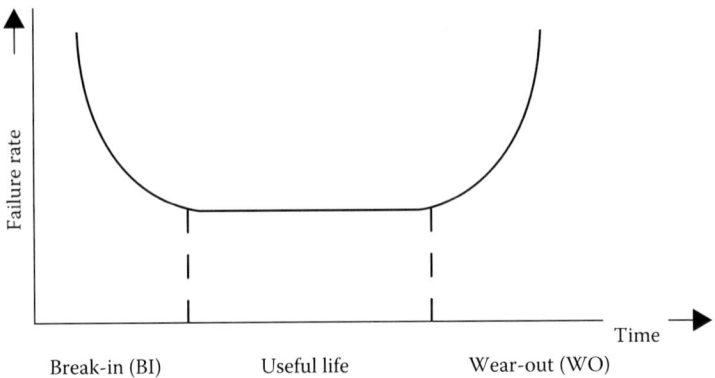

FIGURE 12.1   Bathtub curve.

$$f(t) = \alpha\beta t^{\beta-1} \exp\left[\int_0^t \alpha\beta t^{\beta-1}\, dt\right]$$

$$= \alpha\beta t^{\beta-1} \exp\left(-\alpha t^\beta\right); \quad t > 0; \quad \alpha > 0, \beta > 0 \qquad (12.3)$$

Equation 12.3 primarily defines the pdf of the Weibull distribution. The exponential distribution discussed in the preceding chapter, whose pdf is given in Equation 11.2, is a special case of the Weibull distribution with $\beta = 1$. The variety of assumptions about failure rate and the probability distribution of time to failure that can be accommodated by the Weibull distribution make it especially attractive in describing failure–time distributions in industrial and process plant applications. Estimating the coefficients in the Weibull distribution can be accomplished using a graphical procedure developed by Bury.[3]

To illustrate probability calculations involving the exponential and Weibull distributions introduced in conjunction with the bathtub curve of failure rate, consider first the case of a transistor having a constant rate of failure of 0.01 per thousand hours. To find the probability that the transistor will operate for at least 25,000 h, substitute the failure rate

$$Z(t) = 0.01$$

into Equation 11.24, which yields

$$f(t) = \exp\left[-\int_0^t 0.01\, dt\right]$$

$$= 0.01e^{-0.01t}; \quad t > 0$$

as the pdf of T, the time to failure of the transistor. Because t is measured in thousands of hours, the probability that the transistor will operate for at least 25,000 h is given by

$$P(T > 25) = \int_{25}^{\infty} -0.01e^{-0.01t}\, dt$$

$$= -e^{-\infty} + e^{-0.01(25)}$$

$$= 0 + 0.78$$

$$= 0.78 = 78\%$$

The reader should note that this example reduces to (because of the constant rate specification) a calculation of an exponential distribution.

Now suppose it is desired to determine the 10,000 h reliability of a circuit of five such transistors connected in series. The 10,000 h reliability of one transistor is the probability that it will last at least 10,000 h. This probability can be obtained by integrating the pdf of T, time to failure, which gives

$$P(T > 10) = \int_{10}^{\infty} -0.01e^{-0.01t}\, dt$$

$$= -e^{-\infty} + e^{-0.01(10)}$$

$$= 0 + 0.90$$

$$= 0.90 = 90\%$$

As expected, the result is higher. The same result can also be obtained directly from Equation 11.23, which expresses reliability in terms of failure rate. Substituting the failure rate

$$Z(t) = 0.01$$

into this equation yields

$$R(t) = \exp\left[-\int_{0}^{t} 0.01\, dt \right]$$

as the reliability function. The 10,000 h reliability is therefore (once again)

$$R(10) = \exp\left[-\int_0^{10} 0.01 \, dt\right]$$

$$= e^{-0.01(10)}$$

$$= e^{-0.1}$$

$$= 0.90 = 90\%$$

The 10,000 h reliability of a circuit of five transistors connected in series is obtained by applying the formula for the reliability of series system (see also Equation 11.7):

$$R_s = \left[R(10)\right]^5$$

$$= (0.9)^5$$

$$= 0.59 = 59\%$$

As another example of this type of probability calculation, consider a component whose time to failure T, in hours, has a Weibull pdf with parameters $\alpha=0.01$ and $\beta=0.50$ in Equation 12.3. (Note that this involves a *nonconstant* rate application that applies over the entire time domain.) This gives

$$f(t) = (0.01)(0.50)t^{0.5-1}e^{-(0.01)t^{0.5}}; \quad t > 0$$

$$= (0.005)t^{-0.5}e^{-(0.01)t^{0.5}}; \quad t > 0$$

as the Weibull pdf of the failure time of the component under consideration. The probability that the component will operate at least 8100 h is then given by

$$P(T > 8100) = \int_{8100}^{\infty} f(t) \, dt$$

$$= \int_{8100}^{\infty} 0.005t^{-0.5}e^{-(0.01)t^{0.5}} \, dt$$

$$= e^{-(0.01)t^{0.5}}\Big|_{8100}^{\infty}$$

$$= 0 + e^{-(0.01)(8100)^{0.5}}$$

$$= 0.41 = 41\%$$

A variety of conditional failure distributions, including wear-out patterns, can be accommodated by the Weibull distribution. Therefore, this distribution has been frequently recommended—instead of the exponential distribution—as an appropriate failure distribution model. Empirically satisfactory fits have been obtained from failure data on electron tubes, relays, ball bearings, metal fatigue, and even human mortality.

THE DRaT MODELS

In recent years, the Weibull distribution has come under fire, due primarily to the efforts of the author. The last 3 years has provided an opportunity for Dupont and Theodore, along with Ricci and others,[4–7] to carefully analyze the merits and limitations of the Weibull distribution. This has led to the development of the DRaT models.[8,9] The details are given in the following.

As noted in the previous section, the general two-coefficient Weibull model is represented by an equation that can be applied to three failure rate periods representing three failure mode stages. As such, the model consists of six coefficients—two for each of the three failure mode stages. The author viewed this six-coefficient relationship as both cumbersome *and* unnecessary. After some deliberation, it was decided to employ a new, simpler approach to represent failure behavior—specifically the failure–time (as opposed to failure rate–time) relationship depicted in Figure 12.2. After even more deliberation and analysis, the author settled on four models that are based on the melding of either a power function (P) or an exponential relationship (E) for the BI period with either a power function or exponential term for the WO period—the sum of which results in a curve as shown in Figure 12.2. There are therefore four combinations of the power function and exponential term equations: BI(P)–WO(P), BI(E)–WO(E), BI(E)–WO(P), and BI(P)–WO(E), and the corresponding equations resulting from these combinations are defined as the DRaT II, DRaT III, DRaT IV, and DRaT V models, respectively.

Note that the general relationship for the power function takes the form

$$P(t) = P_1 + P_2 t + P_3 t^2 + P_4 t^3 \tag{12.4}$$

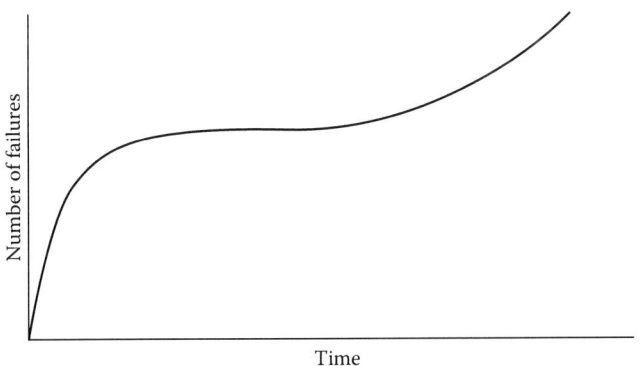

FIGURE 12.2   Failure–time relationship for the Weibull distribution.

The general form of the exponential function takes the form of either

$$E(t) = A_1\left(1 - e^{-A_2 t}\right)$$ (12.5)

for the BI period and

$$E(t) = B_1\left(e^{B_2 t} - 1\right)$$ (12.6)

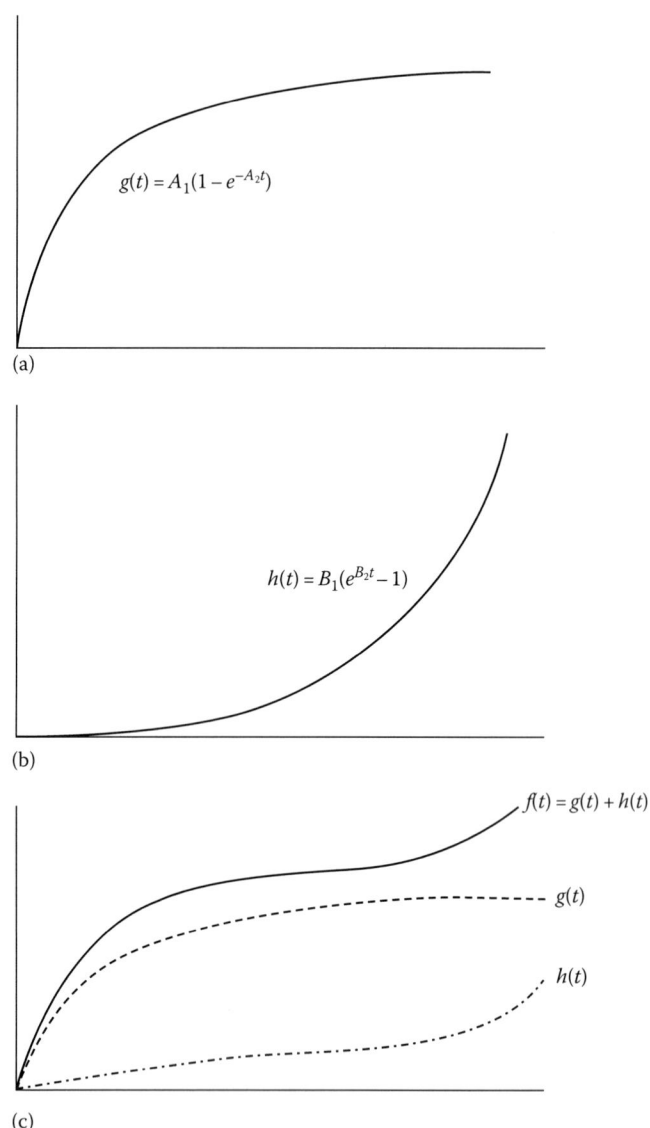

FIGURE 12.3 DRaT model II representation. (a) BI period, (b) WO period, and (c) combined BI and WO period.

for the WO stage. The failure–time functional relationships for the BI and WO periods are represented by $g(t)$ and $h(t)$, respectively, for development to follow, with the sum of the two resulting in the DRaT model, $f(t)$, that is,

$$f(t) = g(t) + h(t) \tag{12.7}$$

For example, if both the BI and WO periods are represented by an exponential function (see also Figure 12.3a,b)

$$g(t) = A_1\left(1 - e^{-A_2 t}\right) \tag{12.8}$$

and

$$h(t) = B_1\left(e^{B_2 t} - 1\right) \tag{12.9}$$

The sum of Equations 12.8 and 12.9, presented in Figure 12.3a, represents the failure model previously defined as the DRaT III model, that is,

$$N = g(t) + h(t) = A_1\left(1 - e^{-A_2 t}\right) + B_1\left(e^{B_2 t} - 1\right); \quad N = \text{number of failures} \tag{12.10}$$

Specific details on each of the DRaT models are provided in the next four sections.

DRaT II Model

This model assumes that BI(P) and WO(P) relationships apply. Thus,

$$N(\text{II}) = N(\text{BI}) + N(\text{WO})$$
$$N(\text{II}) = A_1 + A_2 t + A_3 t^2 + A_4 t^3; \quad A_1 = 0 \tag{12.11}$$

Interestingly, this model only contains three coefficients.

DRaT III Model

This model assumes that BI(E) and WO(E) relationships apply (see also Equation 12.10). Thus,

$$N(\text{BI}) = A_1\left(1 - e^{-A_2 t}\right) \tag{12.12}$$

and

$$N(\text{WO}) = B_1\left(e^{B_2 t} - 1\right) \tag{12.13}$$

so that

$$N\left(\text{III}\right) = A_1\left(1 - e^{-A_2 t}\right) + B_1\left(e^{B_2 t} - 1\right)$$
(12.14)

Note that his model contains four coefficients.

DRaT IV MODEL

This model assumes that BI(E) and WO(P) relationships apply; thus,

$$N\left(\text{BI}\right) = A_1\left(1 - e^{-A_2 t}\right)$$
(12.15)

and

$$N\left(\text{WO}\right) = B_1 + B_2 t + B_3 t^2; \quad B_1 = 0$$
(12.16)

so that

$$N\left(\text{IV}\right) = A_1\left(1 - e^{-A_2 t}\right) + B_2 t + B_3 t^2$$
(12.17)

This model contains four coefficients.

DRaT V MODEL

The last model assumes BI(P) and WO(E) relationships to apply. Thus,

$$N\left(\text{BI}\right) = A_1 + A_2 t + A_3 t^2; \quad A_1 = 0$$
(12.18)

and

$$N\left(\text{WO}\right) = B_1\left(e^{B_2 t} - 1\right)$$
(12.19)

so that

$$N\left(\text{V}\right) = A_2 t + A_3 t^2 + B_1\left(e^{B_2 t} - 1\right)$$
(12.20)

Similar to the last two models, this model also contains four coefficients.

Three improvements in the proposed DRaT models relative to the Weibull model become immediately apparent.

1. The DRaT model requires either three or four (not six) coefficients to be estimated.
2. The DRaT model is continuous over the entire time range as opposed to the Weibull model that is evaluated separately over three compartmentalized failure stages.
3. The requirement of a constant failure rate period in the Weibull model has been replaced by a more realistic failure rate that can continue to slightly increase with time during the supposed *constant* failure rate period.

The reader should note once again that the failure rate for most applications is a *calculated* quantity obtained from the number of failures (N) versus time (t) data. The failure rate at a specified time is approximately equal to the slope (or derivative) of N versus t at the time point in question. (Theodore and Ricci[10] provide six different numerical differentiation procedures to calculate this derivative, and the reader is directed to that reference for more details.)

APPLYING THE DRaT MODELS TO PREDICT FAILURE RATES

As noted earlier, the failure rate (FR) can be calculated if failure (N)–time (t) data is available. The FR is then

$$FR = \frac{dN}{dt} = \frac{d\left[N(t)\right]}{dt} \tag{12.21}$$

The calculations for FR are simplified if dN/dt can be successfully accomplished analytically. Fortunately, the four DRaT models are amenable to analytical differentiation. This derivative, representing the failure rate FR, is provided below for each of the four DRaT models.

DRaT II—the BI(P) and WO(P) model—with N(II) obtained from Equation 12.11:

$$FR(\text{II}) = \left(\frac{dN}{dt}\right)_{\text{II}} = \frac{d\left[A_1 + A_2 t + A_3 t^2 + A_4 t^3\right]}{dt}$$

$$= A_2 + 2A_3 t + 3A_4 t^2 \tag{12.22}$$

DRaT III—the BI(E) and WO(E) model—with N(III) obtained from Equation 12.14:

$$FR(\text{III}) = \left(\frac{dN}{dt}\right)_{\text{III}} = \frac{d\left[A_1\left(1 - e^{-A_2 t}\right) + B_1\left(e^{B_2 t} - 1\right)\right]}{dt}$$

$$= A_1 A_2 e^{-A_2 t} + B_1 B_2 e^{B_2 t} \tag{12.23}$$

DRaT IV—the BI(E) and WO(P) model—with N(IV) obtained from Equation 12.17:

$$FR(IV) = \left(\frac{dN}{dt}\right)_{IV} = \frac{d\left[A_1\left(1 - e^{-A_2 t}\right) + B_2 t + B_3 t^2\right]}{dt}$$

$$= A_1 A_2 e^{-A_2 t} + B_2 + 2B_3 t \tag{12.24}$$

DRaT V—the BI(P) and WO(E) model—with N(V) obtained from Equation 12.20:

$$FR(V) = \left(\frac{dN}{dt}\right)_{V} = \frac{d\left[A_2 t + A_3 t^2 + B_1\left(e^{B_2 t} - 1\right)\right]}{dt}$$

$$= A_2 + 2A_3 t + B_1 B_2 e^{B_2 t} \tag{12.25}$$

APPLYING THE DRaT MODEL TO PREDICT FAILURE INFORMATION

The aforementioned DRaT models can be employed to predict the time to failure of a system by recognizing that the equations are first-order ordinary differential equations describing failure rate as a function of time. For example, separation of variables for the DRaT II model leads to an equation where the limits of integration are from i to n, where i is set to the last recorded data point at which a failure has occurred, and n is the point at which the next failure(s) is(are) to be predicted. Thus, the equation can be rewritten to predict the number of failures, N_2 at t_2, based on the time, t_1 to the number of failures N_1, as

$$N_2 = N_1 + A_2\left(t_2 - t_1\right) + A_3\left(t_2^2 - t_1^2\right) + A_4\left(t_2^3 - t_1^3\right) \tag{12.26}$$

Equation 12.26 can therefore then be solved for time, t_2 after N_2 failures, by any suitable numerical or trial-and-error method provided N_1, N_2, and t_1 are specified. Details of this solution procedure are presented in the last Illustrative Example. DRaT III, DRaT IV, and DRaT V models are provided in Equations 12.27 through 12.29, respectively.

$$N_2 = N_1 - A_1 e^{-A_2(t_2 - t_1)} + B_1 e^{B_2(t_2 - t_1)} \tag{12.27}$$

$$N_2 = N_1 - A_1 e^{-A_2(t_2 - t_1)} + B_2\left(t_2 - t_1\right) + B_3\left(t_2^2 - t_1^2\right) \tag{12.28}$$

$$N_2 = N_1 + A_2\left(t_2 - t_1\right) + A_3\left(t_2^2 - t_1^2\right) + B_1 e^{B_2(t_2 - t_1)} \tag{12.29}$$

ILLUSTRATIVE EXAMPLES

Illustrative Example 12.1

The life (time to failure) of a machine component has a Weibull distribution. Outline how to determine the probability that the component lasts a given period of time if the failure rate is $t^{-1/2}$.

Solution
Identify the failure rate, $Z(t)$, from Equation 12.2,

$$Z(t) = t^{-1/2}$$

Also, identify the values of and appearing in the failure rate equation. If the failure rate is $t^{-1/2}$,

$$\beta - 1 = -1/2$$

and

$$\alpha\beta = 1$$

Therefore, $\beta = 1/2$ and $\alpha = 2$.
    For these values of α and β specified, determine the Weibull pdf:

$$f(t) = t^{-1/2}e^{-2t^{1/2}}; \quad t > 0$$

Integration of this pdf will yield the required probability. This integration is detailed in the next illustrative example.

Illustrative Example 12.2

Refer to the previous example. Determine the probability that the component lasts at least 25,000 h if t is measured in thousands of hours and the failure-rate equation applies over the entire time domain.

Solution
For this case,

$$t = 25$$

Because time is measured in thousands of hours, the probability that the component lasts at least 25,000 h is

$$P(T > 25) = \int_{25}^{\infty} t^{-1/2}e^{-2t^{1/2}}\,dt$$

$$= e^{-2\sqrt{t}}\Big|_{25}^{\infty}$$

This may be integrated to give

$$P(T > 25) = -0 - \left(e^{-2(25)^{0.5}}\right)$$

$$= 4.5 \times 10^5 \text{ h}$$

Illustrative Example 12.3

Refer to the previous example. Obtain the probability equation if the failure rate is given by $t^{-0.75}$.

Solution
For this case,

$$\beta - 1 = -0.25$$
$$\beta = 0.75$$

and

$$\alpha\beta = 1$$
$$\alpha(0.75) = 1$$
$$\alpha = 1.33$$

These values can be substituted back into Equation 12.3.

Illustrative Example 12.4

The life of a gasket has a Weibull distribution with failure rate

$$Z(t) = \frac{1}{\sqrt{t}}$$

where t is measured in years. What is the probability that the gasket will last at least 4 years if the failure rate equation applies across the entire time domain?

Solution
The pdf specified by $f(t)$ in terms of the failure rate, $Z(t)$, is as follows (see Introduction):

$$F(t) = Z(t)\exp\left[-\int_0^t Z(t)dt\right]$$

Substituting $1/t^{1/2}$ for $Z(t)$ yields

$$F(t) = t^{-1/2}\exp\left[-\int_0^t t^{-1/2}dt\right]$$

$$= t^{-1/2}\exp\left(-2t^{1/2}\right)dt; \quad t > 0$$

Employ the integration procedure presented in Illustrative Example 12.2. The probability that the seal lasts at least 4 years is

$$P(T \geq 4) = \int_4^\infty t^{-1/2} e^{-2t^{1/2}} dt$$

$$= e^{-2\sqrt{t}} \Big|_4^\infty$$

$$= -0 - \left(-e^{-4}\right)$$

$$= 0.0183 = 1.83\%$$

Illustrative Example 12.5

The life of an electronic component is a random variable having a Weibull distribution with $\alpha = 0.025$ and $\beta = 0.50$. What is the average life of the component?

Solution

Let T denote the life in hours of the electronic component. The pdf of T is again obtained by applying Equation 12.3, which yields

$$f(t) = \alpha \beta t^{\beta-1} \exp\left(-\alpha t^\beta\right); \quad t > 0; \quad \alpha > 0, \beta > 0$$

Substituting $\alpha = 0.025$ and $\beta = 0.50$ yields

$$f(t) = (0.025)(0.50) t^{0.50-1} \exp\left(-0.025 t^{0.50}\right); \quad t > 0$$

$$= (0.0125) t^{-0.50} \exp\left(-0.025 t^{0.50}\right); 4t > 0$$

The average value for this continuous variable T is given by integration of Equation 4.22:

$$E(T) = \int_{-\infty}^\infty t f(t) dt$$

$$= \int_{-\infty}^\infty (0.0125) t^{-0.50} \exp\left(-0.025 t^{0.50}\right) dt$$

Integrating by any suitable procedure gives

$$E(T) = 3200 \text{ h}$$

Illustrative Example 12.6

Refer to the previous example. What is the probability that the component will last more than 4000 h?

Solution

The probability that the component will last more than 4000 h is given by

$$P(T > 4000) = \int_{4000}^{\infty} f(t)\,dt$$

$$= \int_{4000}^{\infty} (0.0125)t^{-0.50} \exp\left(-0.025t^{0.50}\right)dt$$

$$= e^{-0.025\sqrt{t}}\Big|_{4000}^{\infty}$$

$$= -0 - \left(-e^{-1.581}\right)$$

$$= 0.2057 = 20.57\%$$

Illustrative Example 12.7

Develop an outline for estimating the Weibull parameters.

Solution

Estimating of the parameters in the pdf of the Weibull distribution

$$f(t) = \alpha\beta t^{\beta-1} \exp\left(-\alpha t^{\beta}\right); \quad t > 0; \quad \alpha > 0, \beta > 0$$

can be obtained by a graphical procedure developed by Bury.[3] It is based on the fact that

$$\ln\left[\ln\frac{1}{1-F(t)}\right] = \ln\alpha + \beta\ln t \qquad (12.30)$$

is a linear function of $\ln t$. Here,

$$F(t) = 1 - e^{-\alpha t^{\beta}}; \quad t > 0 \qquad (12.31)$$
$$= 0; \quad t < 0$$

defines the cdf of the Weibull distribution.

The graphical procedure for estimating the slope, β, and the intercept, $\ln \alpha$, of time to failure, involves first ordering the observations from smallest $(i=1)$ to largest $(i=n)$. The value of the ith observation varies from sample to sample. It can be shown that the average value of $F(t)$ for t equal to the value of the ith observation on T is $i/(n+1)$. The points obtained by plotting Equation 12.32 against the natural logarithm of the ith observation for $i=1$ to $i=n$ should lie along a straight line whose slope is β and whose intercept is $\ln \alpha$, if the assumption that T has a Weibull distribution is correct.

$$\ln\left[\ln\frac{1}{1-\dfrac{i}{n+1}}\right] \tag{12.32}$$

This procedure is demonstrated in the next example.

Illustrative Example 12.8

The time in days to failure of each sample of 10 electronic components is observed as follows:

$$71, 40, 90, 149, 127, 53, 106, 36, 18, 165$$

Assuming a Weibull distribution applies, estimate the parameters α and β.

Solution
The observations in order of magnitude are

$$18, 36, 40, 53, 71, 90, 106, 127, 149, 165$$

Table 12.1 is generated from the data. Set $X = \ln t$ and $Y = \ln\{\ln 1/[1 - i/(n+1)]\}$; values of X are in column (3) and values of Y are in column (4).

Compute the values of $\sum_{i=1}^{n} X$. Similarly, calculate $\sum X^2, \sum Y, \sum XY$. These values determine the slop and intercept of the least-squares line of best fit to estimate Y from X. Details of this procedure are provided in the literature.[1,2]

$$\sum X = 42.5; \quad \sum Y = -4.96; \quad n = 10$$

$$\sum X^2 = 185.2; \quad \sum XY = -14.71$$

TABLE 12.1
Weibull Coefficients

(1) Time to Failure (t)	(2) Order of Failure (i)	(3) ln t	(4) ln{ln 1/[1 − i/(n + 1)]}
18	1	2.89	2.35
36	2	3.58	1.61
40	3	3.69	1.14
53	4	3.97	0.79
71	5	4.26	0.50
90	6	4.50	0.24
106	7	4.66	0.01
127	8	4.84	0.26
149	9	5.00	0.53
165	10	5.11	0.87

For this example,

$$\sum_{i=1}^{10}(X)^2 = (2.89)^2 + (3.58)^2 + (3.69)^2 + (3.97)^2 + (4.26)^2 + (4.50)^2$$
$$+ (4.66)^2 + (5.11)^2 + (4.84)^2 + (5.00)^2$$
$$= 185.2$$

Substitute these values in the following formulas for the slope and intercept of the least-squares line of best fit[1,2]:

$$\text{Slope} = \frac{n\sum XY - \left(\sum X\right)\left(\sum Y\right)}{n\sum X^2 - \left(\sum X\right)^2}$$

$$= \frac{(10)(-14.71) - (42.5)(-4.96)}{(10)(185.2) - (42.5)^2}$$

$$= 1.4$$

$$\text{Intercept} = \frac{\sum Y}{n} - (slope)\frac{\sum X}{n}$$

$$= \frac{-4.96}{10} - (1.4)\frac{42.5}{10}$$

$$= -6.4$$

The estimated value of β is 1.4, and the estimated value of α is the antilog of -6.4, which is equal to 0.0017.

REFERENCES

1. L. Theodore and F. Taylor, *Probability and Statistics*, Theodore Tutorials (originally published by USEPA, RTP, NC), East Williston, NY, 1996.
2. S. Shaefer and L. Theodore, *Probability and Statistics Applications in Environmental Science*, CRC Press/Taylor & Francis Group, Boca Raton, FL, 2007.
3. K. Bury, *Statistical Methods in Applied Science*, John Wiley & Sons, Hoboken, NJ, 1975.
4. L. Stander and L. Theodore, Environmental health and hazard risks of nanomaterials, *American Bar Association, Science and Technology Committee Newsletter*, pp. 6–10, Washington, DC, 2010.
5. L. Theodore and A.J Caraccio, Is it a health risk or a hazard risk… or both? Paper #65, *AWMA Conference*, Calgary, CA, 2010.
6. L. Theodore and R. Dupont, Calculating hazard probabilities using the Weibull distribution, Paper #125, *AWMA Conference*, Orlando, FL, 2011.
7. R. Dupont, J. McKenna, and L. Theodore, Baghouse failures: Applying the Weibull distribution to estimate bag failure emissions as a function of time, Paper #12112, *AWMA Conference*, Chicago, IL, 2013.

8. R. Dupont, F. Ricci, and L. Theodore, An improved failure rate model applied to baghouse failures, *AWMA Conference*, Long Beach, CA, 2014.
9. R. Dupont, F. Ricci, and L. Theodore, Replacing the Weibull distribution failure rate model, *AWMA Conference*, Raleigh, NC, 2015.
10. L. Theodore and F. Ricci, *Mass Transfer Operations for the Practicing Engineer*, John Wiley & Sons, Hoboken, NJ, 2011.

13 Other Continuous Probability Distributions

INTRODUCTION

The last four chapters reviewed the key continuous probability distributions that find application in risk assessment/analysis. Although the material primarily addressed on the exponential and Weibull probability distributions, there are numerous applications involving the normal distribution, and to a lesser extent, the log-normal distribution. However, there are also other continuous distributions, the three most important of which are Student's t, χ^2 (chi-square), and the F distributions.

This chapter introduces the reader to the three special distribution indicated earlier, that is,

1. Student's t distribution
2. The χ^2 distribution
3. The F distribution

The next three sections of this chapter will review these three distributions. The chapter concludes with an "all-purpose" section that effectively lists some of the lesser known and rarely used distributions.

STUDENT'S t DISTRIBUTION

The sample mean, $\bar{X}$, constitutes a so-called point estimate of the mean, μ, of the population from which a sample was selected at random. Instead of a *point* estimate, an *interval* estimate of μ may be required along with an indication of the confidence that can be associated with the interval estimate. Such an interval estimate is called a *confidence interval*, and the associated confidence is indicated by a *confidence coefficient*. The length of the confidence interval varies directly with the confidence coefficient for fixed values of n, the sample size; the larger the value of n, the shorter is the confidence interval. Thus, for fixed values of the confidence coefficient, the limits that contain a parameter with a probability of 95% (or some other stated percentage) are defined as the 95% (or that other percentage) *confidence limits* for the parameter; the interval between the confidence limits is referred to as the *confidence interval*. The reader is referred to Chapter 9 for additional details.

For normal distributions, the confidence coefficient Z can be obtained from the standard normal table for various confidence limits or corresponding levels of significance for μ. Some of these values are provided in Table 13.1. Thus, this table can be employed to obtain the probability of a value falling inside or outside the

TABLE 13.1

Confidence Levels and Levels of Significance

Confidence Level (%)	Level of Significance (%)	Z
80.00	20.00	1.28
90.00	10.00	1.65
95.00	5.00	1.96
95.45	4.55	2.00
98.00	2.00	2.33
99.00	1.00	2.58
99.73	0.27	3.00
99.90	0.10	3.29
99.99	0.01	3.89

range of $\mu \pm Z\sigma$. For example, when $Z = 2.58$, one can say that the level of significance is 1%. The statistical interpretation of this is as follows: if an observation deviates from the mean by at least ± 2.58, σ, the observation is significantly different from the body of data on which the describing normal distribution is based. Further, the probability that this statement is in error is 1%; that is, the conclusion drawn from a rejected observation will be wrong 1% of the time.

From Table 13.1, one can also state that there is a 99% probability that an observation will fall within the range $\mu \pm 2.58$. The degree of confidence is referred to as the 99% *confidence level*. The limits, $\mu - 2.58\sigma$ and $\mu + 2.58\sigma$, are defined as the *confidence limits*, whereas the difference between the two values is defined as the *confidence interval*. Once again, the example in the preceding text essentially states that the actual (or true) mean lies within the interval $\mu - 2.58\sigma$ and $\mu + 2.58\sigma$ with a 99% probability of being correct.

The foregoing analysis can be extended to provide the difference of two population means, that is, $\bar{X}_2 - \bar{X}_1$. For this case, the confidence limits are

$$\bar{X}_2 - \bar{X}_1 \pm Z\sqrt{\frac{\sigma_2^2}{n_2} + \frac{\sigma_1^2}{n_1}} \tag{13.1}$$

This analysis set also serves to introduce Student's t distribution. It is common to use this distribution if the sample size is *small*. For a random sample of size n selected from a normal population, the term $(\bar{X} - \mu)/(s - \sqrt{n})$ has Student's distribution with $(n - 1)$ degrees of freedom. *Degrees of freedom* is the label used for the parameter appearing in the Student's distribution pdf in Equation 13.2 (Figure 13.1).

$$f(t) = \frac{\Gamma\left(\frac{v+1}{2}\right)}{\sqrt{\pi v}\,\Gamma\left(\frac{v}{2}\right)}\left(1 + \frac{t^2}{v}\right)^{\frac{v+1}{2}} \; ; \quad -\infty < t < \infty; \quad \Gamma = \text{Gamma function} \tag{13.2}$$

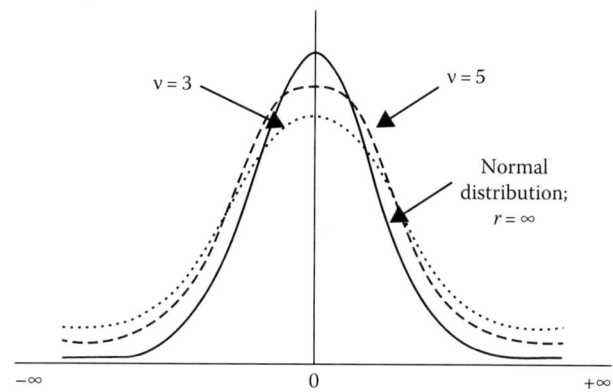

FIGURE 13.1   Student's t distribution.

The graph of the pdf of Student's distribution is symmetric about 0 for all values of the parameter. The graph is similar to but not as peaked as the standard normal curve presented earlier in Chapter 9.

Once again, assume X is normally distributed with mean μ and variance σ^2. Let $\bar{X}$ and s^2 represent the corresponding sample estimates on 10 random samples. For this case, Z is still given by

$$Z = \frac{\bar{X} - \mu}{\sigma/\sqrt{10}} \tag{13.3}$$

$$t = \frac{\bar{X} - \mu}{s/\sqrt{10}} \tag{13.4}$$

where t is drawn from Student's t distribution with $n - 1$ degrees of freedom (usually designated as v, where $v = n$). If $t_{0.01}$ represents the value of t such that the probability is 1% that $|t| > t_{0.01}$, one can conclude that the probability is 99% that

$$\frac{\bar{X} - \mu}{s/\sqrt{10}} < t_{0.01}$$

The preceding equation may also be rewritten as

$$\bar{X} - t_{0.01}\frac{s}{\sqrt{10}} < \mu < \bar{X} + t_{0.01}\frac{s}{\sqrt{10}}$$

For this *two-sided* test, the appropriate value of t from Table 13.2 is 3.25 (probability of 0.005 for each tail, with $r = 9$):

$$\bar{X} - \frac{3.25s}{\sqrt{10}} < \mu < \bar{X} + \frac{3.25s}{\sqrt{10}} = 0.99$$

TABLE 13.2
Student's _t_ Distribution

v	α = 0.10	α = 0.05	α = 0.025	α = 0.01	α = 0.005	v
1	3.077684	6.313752	12.70620	31.82052	63.65674	1
2	1.885618	2.919986	4.30265	6.96456	9.92484	2
3	1.637744	2.353363	3.18245	4.54070	5.84091	3
4	1.533206	2.131847	7.77645	3.74695	4.60409	4
5	1.475884	2.015048	2.57058	3.36493	4.03214	5
6	1.439756	1.943180	2.44691	3.14267	3.70743	6
7	1.414924	1.894579	2.36462	2.99795	3.49948	7
8	1.396815	1.859548	2.30600	2.89646	3.35539	8
9	1.383029	1.833113	2.26216	2.82144	3.24984	9
10	1.372184	1.812461	2.22814	2.76377	3.16927	10
11	1.363430	1.795885	2.20099	2.71808	3.10581	11
12	1.356217	1.782288	2.17881	2.68100	3.05454	12
13	1.350171	1.770933	2.16037	2.65031	3.01228	13
14	1.345030	1.761310	2.14479	2.62449	2.97684	14
15	1.340606	1.753050	2.13145	2.60248	2.94671	15
16	1.336757	1.745884	2.11991	2.58349	2.92078	16
17	1.333379	1.739607	2.10982	2.56693	2.89823	17
18	1.330391	1.734064	2.10092	2.55238	2.87844	18
19	1.327729	1.729133	2.09302	2.53948	2.86093	19
20	1.325341	1.724718	2.08596	2.52798	2.84534	20
22	1.321237	1.717144	2.07387	2.50832	2.81876	22
24	1.317836	1.710882	2.06390	2.49216	2.79694	24
26	1.314972	1.705618	2.05553	2.47863	2.77871	26
28	1.312527	1.701131	2.04841	2.46714	2.76326	28
∞	1.281552	1.644854	1.95996	2.32635	2.57583	∞

Source: http://www.statsoft.com/textbook/stable.html.

This may also be written as

$$P(-3.25 < T < 3.25) = 0.99 = 99\%$$

It should be noted that the procedure given earlier provides confidence limit information on both sides of the mean, often referred to as the two _tails_ of the interval. For information on _one side_ of the tail (or a "one-tailed" test), the interval is located on one side of the mean, with the level of significance totally associated with that side. Applications include testing whether one mean is better than another, as opposed to whether it is worse.

Consider the following health risk problem. An environmental scientist has been informed that the standard deviation of a toxic impurity in a nanochemical is 0.23%. The scientist later drew 36 samples from a new batch of the chemical, and the average impurity was 1.92%. Calculate the 99% confidence interval for the true mean. Assume that $\overline{X}$ is normally distributed. The standard normal variable is

$$Z = \frac{\overline{X} - \mu}{\sigma/\sqrt{n}} \tag{13.3}$$

Noting that

$$\overline{X} = 1.92$$

$$\sigma = 0.23$$

$$n = 36$$

one obtains

$$Z = \frac{1.92 - \mu}{0.038}$$

From Table 13.1, one determines that the 99% two-tailed probability is approximately between −2.57 and +2.57. Therefore,

$$P\left(-2.57 < \frac{\overline{X} - \mu}{0.038} < 2.57\right) = 0.99$$

$$P\left(-0.0985 < \overline{X} - \mu < 0.0985\right) = 0.99$$

Thus, the probability is approximately 99% that $\overline{X}$ will be within 0.0985 of the true mean μ. With the observed value of $\overline{X} = 1.92$, one can say there is a 99% confidence that the true mean is in the interval 1.92 ± 0.0985. Thus,

$$1.822 < \mu < 2.019$$

Summarizing, the Student's t distribution is employed when dealing with small samples. The graph of the pdf of Student's distribution is symmetric about 0 for all values of the parameter. The graphs are similar to but not as peaked as the standard normal curve (see Chapter 4). The graph approaches the normal distribution in the limit when the degrees of freedom approach infinity. The term t_ν has also been used to designate a random variable having Student's distribution with ν degrees of freedom. When the sample size is small ($n < 30$), σ is unknown, and the population sample is assumed to be normal, the statistic $(\overline{X} - \mu)/(s - \sqrt{n})$ can, as described earlier, be used to generate confidence intervals for the population mean μ. As noted in Chapter 9, $\overline{X}$ is an estimate of the true mean μ. The t function provides the distribution of deviations of $\overline{X}$ from μ in terms of probabilities (or relative frequencies).

If one rewrites the t equation in terms of plus or minus, the true mean for this class is given by

$$\mu = \bar{X} \pm ts \tag{13.5}$$

In ordinary language, the true mean can be said, with the tabulated probability of error, to be within the range of the calculated mean included in the limits of plus and minus t times the estimated standard deviation of the mean. It should be once again noted that one is merely applying the confidence interval numbers to a given experiment. It is obviously incorrect to state that the probability is 0.95 (or 0.99, etc.) and that the interval contains μ; the latter probability is either 1 or 0, depending on whether μ does or does not lie in this interval. What it does mean is that μ will lie within the interval 95 out of 100 times. It is only when the random interval $\bar{X} \pm ts$ is considered that one can make correct probability statements of the type desired.

CHI-SQUARE (χ^2) DISTRIBUTION

A random variable X is said to have a chi-square distribution with v degrees of freedom if its probability distribution function (pdf) is specified by

$$f\left(\chi^2\right) = \frac{1}{2^{\frac{v}{2}}\Gamma\left(\frac{v}{2}\right)}\left(\chi^2\right)^{\frac{v}{2}-1} e^{\frac{-\chi^2}{2}}; \quad \chi^2 > 0; \quad \Gamma = \text{Gamma function} \tag{13.6}$$

or

$$f\left(\chi\right) = \frac{1}{2^{\frac{v}{2}}\Gamma\left(\frac{v}{2}\right)}\left(\chi\right)^{\frac{v}{2}-1} e^{\frac{-\chi}{2}}; \quad \chi > 0; \quad \Gamma = \text{Gamma function} \tag{13.7}$$

where
    χ is the Greek letter chi (chi is pronounced as in *kite*)
    χ^2 is read as chi-square

The aforementioned equation may also be written as

$$f\left(\chi^2\right) = A\left(\chi^2\right)^{\frac{v}{2}-1} e^{\frac{-\chi^2}{2}} \tag{13.8}$$

where
    $v = n - 1$ is once again the number of degrees of freedom
    A is a constant depending on a value of v such that the total area under the curve
        is unity

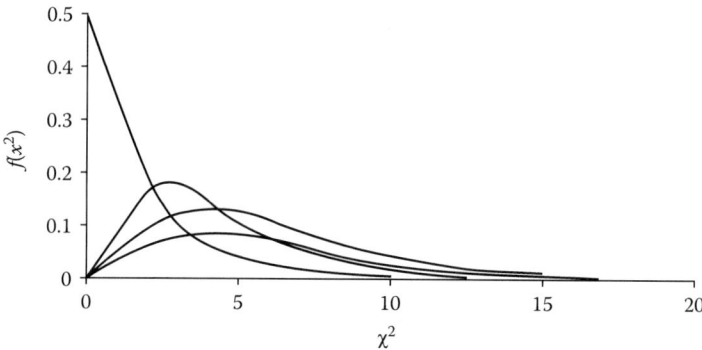

FIGURE 13.2 Chi-square distributions for various values of v.

The chi-square distributions corresponding to various values of v are shown in Figure 13.2. The maximum value of $f(\chi^2)$ occurs at $\chi^2 = v - 2$ for $v \geq 2$.

For random samples from a normal population with variance σ^2, the statistic

$$\chi^2 = \frac{(n-1)s^2}{\sigma^2} \tag{13.9}$$

where s^2 is the sample variance, and n, the sample size, has a chi-square distribution with $n - 1$ degrees of freedom. The variance may be calculated using the procedures outlines earlier.

A test of the hypothesis H_0: $\sigma^2 = \sigma_0^2$[1] utilizes the test statistic

$$\frac{(n-1)s^2}{\sigma_0^2} \tag{13.9}$$

The critical region is determined by the alternative hypothesis, H_1, and α, the tolerated probability of a Type I error.[1,2]

As with the normal and t distributions, one can define 95%, 99%, or other confidence limits and intervals for χ^2 by use of the table of the χ^2 distribution (see Table 13.3). In this manner, one can estimate within specified limits of confidence the population standard deviation σ in terms of a sample standard deviation, s. For example, $\chi_{.025}^2$ and $\chi_{.975}^2$ are the values of χ^2 (called *critical values*), for which 2.5% of the area lies in each "tail" of the distribution. The 95% confidence interval is then

$$\chi_{.025}^2 < \frac{(n-1)s^2}{\sigma^2} < \chi_{0.975}^2$$

from which one can see that σ is estimated to lie in the interval

$$\frac{s\sqrt{n-1}}{\chi_{0.975}} < \sigma < \frac{s\sqrt{n-1}}{\chi_{.025}}$$

TABLE 13.3
Chi-Square Distribution

Values of $x_{\alpha,v}^2$

v	$\alpha=0.995$	$\alpha=0.990$	$\alpha=0.975$	$\alpha=0.950$	$\alpha=0.900$	$\alpha=0.750$	v
1	0.000	0.000	0.001	0.004	0.016	0.102	1
2	0.010	0.020	0.051	0.103	0.211	0.575	2
3	0.071	0.115	0.216	0.352	0.584	1.21	3
4	0.207	0.297	0.484	0.711	1.06	1.92	4
5	0.412	0.554	0.831	1.15	1.61	2.67	5
6	0.676	0.872	1.24	1.64	2.20	3.45	6
7	0.990	1.24	1.69	2.17	2.83	4.25	7
8	1.34	1.65	2.18	2.73	3.49	5.07	8
9	1.73	2.09	2.70	3.33	4.17	5.90	9
10	2.16	2.56	3.25	3.94	4.87	6.74	10
11	2.60	3.05	3.82	4.57	5.58	7.58	11
12	3.07	3.57	4.40	5.23	6.30	8.44	12
13	3.57	4.11	5.01	5.89	7.04	9.30	13
14	4.07	4.66	5.63	6.57	7.79	10.2	14
15	4.60	5.23	6.26	7.26	8.55	11.0	15
16	5.14	5.81	6.91	7.96	9.31	11.9	16
17	5.70	6.41	7.56	8.67	10.1	12.8	17
18	6.26	7.01	8.23	9.39	10.9	13.7	18
19	6.84	7.63	8.91	10.1	11.7	14.6	19
20	7.43	8.26	9.59	10.9	12.4	15.5	20
22	8.64	9.54	11.0	12.3	14.0	17.2	22
24	9.89	10.9	12.4	13.8	15.7	19.0	24
26	11.2	12.2	13.8	15.4	17.3	20.8	26
28	12.5	13.6	15.3	16.9	18.9	22.7	28
30	13.8	15.0	16.8	18.5	20.6	24.5	30

Values of $x_{\alpha,v}^2$

v	$\alpha=0.50$	$\alpha=0.25$	$\alpha=0.10$	$\alpha=0.05$	$\alpha=0.025$	$\alpha=0.01$	$\alpha=0.005$	v
1	.455	1.32	2.71	3.84	5.02	6.63	7.88	1
2	1.39	2.77	4.61	5.99	7.38	9.21	10.7	2
3	2.37	4.11	6.25	7.81	9.35	11.3	12.8	3
4	3.36	5.39	7.78	9.49	11.1	13.3	14.9	4
5	4.35	6.63	9.24	11.1	12.8	15.1	16.7	5
6	5.35	7.84	10.6	12.6	14.4	16.8	18.5	6
7	6.35	9.04	12.0	14.1	16.0	18.5	20.3	7
8	7.34	10.2	13.4	15.5	17.5	20.1	22.0	8
9	8.34	11.4	14.7	16.9	19.0	21.7	23.6	9
10	9.34	12.5	16.0	18.3	20.5	23.2	25.2	10
11	10.3	13.7	17.3	19.7	21.9	24.7	26.8	11
12	11.3	14.8	18.5	21.0	23.3	26.2	28.3	12

(*Continued*)

TABLE 13.3 (*Continued*)
Chi-Square Distribution

Values of $x_{\alpha,v}^2$

v	$\alpha=0.50$	$\alpha=0.25$	$\alpha=0.10$	$\alpha=0.05$	$\alpha=0.025$	$\alpha=0.01$	$\alpha=0.005$	v
13	12.3	16.0	19.8	22.4	24.7	27.7	29.8	13
14	13.3	17.1	21.1	23.7	26.1	29.1	31.3	14
15	16.3	18.2	22.3	25.0	27.5	30.6	32.8	15
16	15.3	19.4	23.5	26.3	28.8	32.0	34.3	16
17	16.3	20.5	24.8	27.6	30.2	33.4	35.7	17
18	17.3	21.6	26.0	28.9	31.5	34.8	37.2	18
19	18.3	22.7	27.2	30.1	32.9	36.2	38.6	19
20	19.3	23.8	28.4	31.4	34.2	37.6	40.0	20
22	21.3	26.0	30.8	33.9	36.8	40.3	42.8	22
24	23.3	28.2	33.2	36.4	39.4	43.0	45.6	24
26	25.3	30.4	35.6	38.9	41.9	45.6	48.3	26
28	27.3	32.6	37.9	41.3	44.5	48.3	51.0	28
30	29.3	34.8	40.3	43.8	47.0	50.9	53.7	30

Source: http://www.statsoft.com/textbook/stable.html.

with 95% confidence. Similarly, other confidence intervals can be found. The values $\chi_{.025}$ and $\chi_{0.975}$ represent, respectively, the 2.5 and 97.5 percentile values. The preceding two equations may more appropriately be written as

$$\chi_{.025}^2 < \frac{(n-1)s^2}{\sigma^2} \quad \text{or} \quad \chi_{0.975}^2 > \frac{(n-1)s^2}{\sigma^2}$$

and

$$\sigma < \frac{s\sqrt{n-1}}{\chi_{.025}} \quad \text{or} \quad \sigma > \frac{s\sqrt{n-1}}{\chi_{0.975}}$$

Table 13.3 gives percentile values corresponding to the number of degrees of freedom. For large values of v ($v \geq 30$), one can use the fact that $\left(\sqrt{2\chi^2} - \sqrt{2v-1}\right)$ is very nearly normally distributed with mean 0 and standard deviation 1 so that normal distribution tables can be used (if $v \geq 30$). Then if χ_p^2 and Z_p are the Pth percentiles of the chi-square and normal distributions, respectively, one has

$$\chi_p^2 = \frac{1}{2}\left(Z_p + \sqrt{2v+1}\right)^2 \tag{13.11}$$

The chi-square distribution is of great practical and theoretical interest. It is used in the following areas:

1. To calculate a confidence interval from sample data for the variance of a normal distribution
2. To calculate a confidence interval from sample data for the parameter of an exponential distribution
3. To conduct various "goodness-of-fit" tests to determine whether a given sample could reasonably have come from a normal distribution
4. To determine whether significant differences exist among the frequencies in a contingency table

Occasionally, the chi-square distribution has provided a reasonable approximation of the distribution of certain physical variables. For example, it has represented compressor failure of large jet engines by a chi-square distribution with 5 degrees of freedom.
Consider the following two questions:

1. For 2 degrees of freedom ($v=2$), what is the probability of obtaining a χ^2 value equal to or greater than 4.605?
2. For 20 degrees of freedom, what value of χ^2 will provide a 5% (0.05) probability that χ^2 is equal to or larger than this value?

From Table 13.3,

1. Probability $=0.10=10\%$
2. $\chi^2=31.41$

Consider now a chi-square distribution with 9 degrees of freedom. Find the critical values for $\chi^2_{0.05}$ and $\chi^2_{0.95}$. Also obtain the critical values for $\chi^2_{0.05}$ and $\chi^2_{0.95}$ if the degrees of freedom is 14. Refer to Table 13.3. For $v=9$,

$$\chi^2_{0.05} = 16.919$$

$$\chi^2_{0.95} = 3.325$$

For $v=14$,

$$\chi^2_{0.05} = 23.685$$

$$\chi^2_{0.95} = 6.571$$

A similar procedure could be employed to obtain the 90% confidence interval for χ^2 if $\alpha=0.05$ with $v=19$. For this case, refer to Equation 13.9.

$$\chi^2_{0.95} < \chi^2 < \chi^2_{0.05}$$

$$\chi^2_{0.95} < \frac{(n-1)s^2}{\sigma^2} < \chi^2_{0.05}$$

If the sample standard deviation, s, was specified, the confidence interval for the population standard deviation (inverting the above equation) would be given by

$$\frac{s\sqrt{n-1}}{\sqrt{30.1}} < \sigma < \frac{s\sqrt{n-1}}{\sqrt{10.1}}$$

Now consider the following health risk study associated with auto emissions. A car manufacturer reports to the U.S. Environmental Protection Agency (USEPA) that the gas mileage in miles per gallon (mpg) for a new model has a mean equal to 38 and a standard deviation equal to 3. A consumer service tests a random sample of 15 cars of the new model and obtains the following gas mileages:

37 39 42 45 34 32 36 36 38 43 40 43 37 30 38

Assuming the gas mileage is normally distributed, test the hypothesis H_0: $\sigma = 3$ against the alternative hypothesis H_1: $\sigma > 3$. Use 0.05 as the tolerated probability of a Type I error. Note that the gas mileage is assumed to be normally distributed. The one-tail critical regional statistic is

$$\frac{(n-1)s^2}{\sigma_0^2} > \chi_{0.05}^2$$

For H_0: $\sigma = 3$, the value of σ_0 is 3. From the table of the chi-square distribution, $\chi_{0.05}^2 = 23.7$ for $v = 14$. Therefore, the critical region is specified by

$$\frac{(n-1)s^2}{9} > 23.7$$

The value of the test statistic $(n-1)s^2/\sigma_0^2$ may now be calculated. For the given data, $n = 15$ and, as can be calculated using the procedures set forth in Chapter 3, $s = 4.19$. Therefore, the value of the test statistic is

$$\frac{(14)(4.19)^2}{9} = 27.3$$

The hypothesis test may now be completed. Reject H_0: $\sigma = 3$, because the observed value of the test statistic, 27.3, exceeds 23.7. Thus, the data do not support the manufacturer's claim that the standard deviation of gas mileage for the new model is 3 mpg. The standard deviation exceeds 3 mpg.

F DISTRIBUTION

As noted several times earlier, variance measures variability. Comparison of variability, therefore, involves comparison of variance. Suppose σ_1^2 and σ_2^2 represent the unknown variances of two independent normal populations. The null hypothesis, H_0: $\sigma_1^2 = \sigma_2^2$, asserts that the two populations have the same variance and are therefore characterized by the same variability. When the null hypothesis is true, the test statistic, s_1^2/s_2^2, has an F distribution with parameters $n_1 - 1$ and $n_2 - 1$ degrees of freedom. Here, s_1^2 is the sample variance of a random sample of n_1 observations from the normal population having variance σ_1^2, while s_2^2 is the sample variance of a random sample n_2 observations from an independent normal population having variance σ_2^2. When s_1^2/s_2^2 is used as a test statistic, the critical region depends on the alternative hypothesis, H_1, and α, the tolerated probability of a Type I error. Listed in Table 13.4 are the critical regions for three different alternative hypotheses.

The term F_α is the value which a random variable having an F distribution with $n_1 - 1$ and $n_2 - 1$ degrees of freedom exceeds with probability α. The terms $F_{\alpha/2}$ and $F_{1-\alpha/2}$ are defined similarly. Also note that

$$F_{\alpha;v_1,v_2} = \frac{1}{F_{1-\alpha;v_2,v_1}} \tag{13.11}$$

where v_1 and v_2 are again the degrees of freedom. Tabulated values of the F distribution are provided in Tables 13.5 and 13.6. In addition, the ratio s_1^2/s_2^2 must, out of necessity, be greater than unity, that is, $s_1^2/s_2^2 > 1.0$.

Student's t distribution was discussed previously. With regard to means, it was noted that samples drawn from the same source or population, or measurements of other types, could not be expected to be identical. Obviously, the means of several samples vary over a range and this range can be approximated for any desired probability level. Similar comments apply for variances. As with means, if several groups of samples are obtained, the calculated variance of the population source obtained from each sample, once again, will not be the same.

The F test and F distribution provide a method for determining whether the ratio of two variances is larger than might be normally expected by chance if the samples came from the same source or population. The procedure to follow is similar to that for comparing two means.

The null hypothesis for comparing two variances is

$$H_0 = \sigma^2(X_1) = \sigma^2(X_2) \tag{13.12}$$

TABLE 13.4

Critical Regions: F Distribution

Alternative Hypothesis (H_1)	Critical Region
$\sigma_1^2 \neq \sigma_2^2$	$s_1^2/s_2^2 < F_{1-\alpha/2}$ or $s_1^2/s_2^2 > F_{\alpha/2}$
$\sigma_1^2 > \sigma_2^2$	$s_1^2/s_2^2 > F_\alpha$
$\sigma_1^2 < \sigma_2^2$	$s_1^2/s_2^2 < F_\alpha$

TABLE 13.5
F Distribution (a)

Values of $F_{0.01, \nu_1, \nu_2}$

ν_2 \ ν_1	1	2	3	4	5	6	7	8	9	10	15	20	30	40	60	120	∞
1	4052.1	4999.5	5403.4	5624.6	5763.7	5859.0	5928.4	5981.1	6022.5	6055.8	6157.3	6208.7	6260.6	6286.8	6313.0	6339.4	6365.9
2	98.5	99.0	99.17	99.25	99.3	99.33	99.36	99.37	99.39	99.4	99.43	99.45	99.47	99.47	99.48	99.49	99.5
3	34.12	30.82	29.46	28.71	28.24	27.91	27.67	27.49	27.35	27.23	26.87	26.69	26.51	26.41	26.32	26.22	26.13
4	21.2	18.0	16.69	15.98	15.52	15.21	14.98	14.8	14.66	14.55	14.2	14.02	13.84	13.75	13.65	13.56	13.46
5	16.26	13.27	12.06	11.39	10.97	10.67	10.46	10.29	10.16	10.05	9.722	9.553	9.379	9.291	9.202	9.112	9.02
6	13.75	10.93	9.78	9.148	8.746	8.466	8.26	8.102	7.976	7.874	7.559	7.396	7.229	7.143	7.057	6.969	6.88
7	12.25	9.547	8.451	7.847	7.46	7.191	6.993	6.84	6.719	6.62	6.314	6.155	5.992	5.908	5.824	5.737	5.65
8	11.26	8.649	7.591	7.006	6.632	6.371	6.178	6.029	5.911	5.814	5.515	5.359	5.198	5.116	5.032	4.946	4.859
9	10.56	8.022	6.992	6.422	6.057	5.802	5.613	5.467	5.351	5.257	4.962	4.808	4.649	4.567	4.483	4.398	4.311
10	10.04	7.559	6.552	5.994	5.636	5.386	5.2	5.057	4.942	4.849	4.558	4.405	4.247	4.165	4.082	3.996	3.909
11	9.646	7.206	6.217	5.668	5.316	5.069	4.886	4.744	4.632	4.539	4.251	4.099	3.941	3.86	3.776	3.69	3.602
12	9.33	6.927	5.953	5.412	5.064	4.821	4.64	4.499	4.388	4.296	4.01	3.858	3.701	3.619	3.535	3.449	3.361
13	9.074	6.701	5.739	5.205	4.862	4.62	4.441	4.302	4.191	4.1	3.815	3.665	3.507	3.425	3.341	3.255	3.165
14	8.862	6.515	5.564	5.035	4.695	4.456	4.278	4.14	4.03	3.939	3.656	3.505	3.348	3.266	3.181	3.094	3.004
15	8.683	6.359	5.417	4.893	4.556	4.318	4.142	4.004	3.895	3.805	3.522	3.372	3.214	3.132	3.047	2.959	2.868
16	8.531	6.226	5.292	4.773	4.437	4.202	4.026	3.89	3.78	3.691	3.409	3.259	3.101	3.018	2.933	2.845	2.753
17	8.4	6.112	5.185	4.669	4.336	4.102	3.927	3.791	3.682	3.593	3.312	3.162	3.003	2.92	2.835	2.746	2.653
18	8.285	6.013	5.092	4.579	4.248	4.015	3.841	3.705	3.597	3.508	3.227	3.077	2.919	2.835	2.749	2.66	2.566
19	8.185	5.926	5.01	4.5	4.171	3.939	3.765	3.631	3.523	3.434	3.153	3.003	2.844	2.761	2.674	2.584	2.489
20	8.096	5.849	4.938	4.431	4.103	3.871	3.699	3.564	3.457	3.368	3.088	2.938	2.778	2.695	2.608	2.517	2.421
21	8.017	5.78	4.874	4.369	4.042	3.812	3.64	3.506	3.398	3.31	3.03	2.88	2.72	2.636	2.548	2.457	2.36

(Continued)

TABLE 13.5 (Continued)
F Distribution (a)

Values of $F_{0.01, \nu_1, \nu_2}$

ν_1

ν_2	1	2	3	4	5	6	7	8	9	10	15	20	30	40	60	120	∞
22	7.945	5.719	4.817	4.313	3.988	3.758	3.587	3.453	3.346	3.258	2.978	2.827	2.667	2.583	2.495	2.403	2.305
23	7.881	5.664	4.765	4.264	3.939	3.71	3.539	3.406	3.299	3.211	2.931	2.781	2.62	2.535	2.447	2.354	2.256
24	7.823	5.614	4.718	4.218	3.895	3.667	3.496	3.363	3.256	3.168	2.889	2.738	2.577	2.492	2.403	2.31	2.211
25	7.77	5.568	4.675	4.177	3.855	3.627	3.457	3.324	3.217	3.129	2.85	2.699	2.538	2.453	2.364	2.27	2.169
26	7.721	5.526	4.637	4.14	3.818	3.591	3.421	3.288	3.182	3.094	2.815	2.664	2.503	2.417	2.327	2.233	2.131
27	7.677	5.488	4.601	4.106	3.785	3.558	3.388	3.256	3.149	3.062	2.783	2.632	2.47	2.384	2.294	2.198	2.097
28	7.636	5.453	4.568	4.074	3.754	3.528	3.358	3.226	3.12	3.032	2.753	2.602	2.44	2.354	2.263	2.167	2.064
29	7.598	5.42	4.538	4.045	3.725	3.499	3.33	3.198	3.092	3.005	2.726	2.574	2.412	2.325	2.234	2.138	2.034
30	7.562	5.39	4.51	4.018	3.699	3.473	3.304	3.173	3.067	2.979	2.7	2.549	2.386	2.299	2.208	2.111	2.006
40	7.314	5.179	4.313	3.828	3.514	3.291	3.124	2.993	2.888	2.801	2.522	2.369	2.203	2.114	2.019	1.917	1.805
60	7.077	4.977	4.126	3.649	3.339	3.119	2.953	2.823	2.718	2.632	2.352	2.198	2.028	1.936	1.836	1.726	1.601
120	6.851	4.787	3.949	3.48	3.174	2.956	2.792	2.663	2.559	2.472	2.192	2.035	1.86	1.763	1.656	1.533	1.381
∞	6.635	4.605	3.782	3.319	3.017	2.802	2.639	2.511	2.407	2.321	2.039	1.878	1.696	1.592	1.473	1.325	1

ν_1 = Degrees of freedom.
ν_2 = Degrees of freedom.

TABLE 13.6
F Distribution (b)

Values of $F_{0.05, \nu_1, \nu_2}$

ν_1

ν_2	1	2	3	4	5	6	7	8	9	10	15	20	30	40	60	120	∞
1	161.4	199.5	215.7	224.6	230.2	234	236.8	238.9	240.5	241.9	245.9	248	250.1	251.1	252.2	253.3	254.3
2	18.51	19	19.16	19.25	19.3	19.33	19.35	19.37	19.38	19.4	19.43	19.45	19.46	19.47	19.48	19.49	19.5
3	10.13	9.552	9.277	9.117	9.014	8.941	8.887	8.845	8.812	8.786	8.703	8.66	8.617	8.594	8.572	8.549	8.526
4	7.709	6.944	6.591	6.388	6.256	6.163	6.094	6.041	5.999	5.964	5.858	5.803	5.746	5.717	5.688	5.658	5.628
5	6.608	5.786	5.41	5.192	5.05	4.95	4.876	4.818	4.773	4.735	4.619	4.558	4.496	4.464	4.431	4.399	4.365
6	5.987	5.143	4.757	4.534	4.387	4.284	4.207	4.147	4.099	4.06	3.938	3.874	3.808	3.774	3.74	3.705	3.669
7	5.591	4.737	4.347	4.12	3.972	3.866	3.787	3.726	3.677	3.637	3.511	3.445	3.376	3.34	3.304	3.267	3.23
8	5.318	4.459	4.066	3.838	3.688	3.581	3.501	3.438	3.388	3.347	3.218	3.15	3.079	3.043	3.005	2.967	2.928
9	5.117	4.257	3.863	3.633	3.482	3.374	3.293	3.23	3.179	3.137	3.006	2.937	2.864	2.826	2.787	2.748	2.707
10	4.965	4.103	3.708	3.478	3.326	3.217	3.136	3.072	3.02	2.978	2.845	2.774	2.7	2.661	2.621	2.58	2.538
11	4.844	3.982	3.587	3.357	3.204	3.095	3.012	2.948	2.896	2.854	2.719	2.646	2.571	2.531	2.49	2.448	2.405
12	4.747	3.885	3.49	3.259	3.106	2.996	2.913	2.849	2.796	2.753	2.617	2.544	2.466	2.426	2.384	2.341	2.296
13	4.667	3.806	3.411	3.179	3.025	2.915	2.832	2.767	2.714	2.671	2.533	2.459	2.38	2.339	2.297	2.252	2.206

(Continued)

TABLE 13.6 (Continued)
F Distribution (b)

Values of $F_{0.05, \nu_1, \nu_2}$

ν_2 \ ν_1	1	2	3	4	5	6	7	8	9	10	15	20	30	40	60	120	∞
14	4.6	3.739	3.344	3.112	2.958	2.848	2.764	2.699	2.646	2.602	2.463	2.388	2.308	2.266	2.223	2.178	2.131
15	4.543	3.682	3.287	3.056	2.901	2.791	2.707	2.641	2.588	2.544	2.403	2.328	2.247	2.204	2.16	2.114	2.066
16	4.494	3.634	3.239	3.007	2.852	2.741	2.657	2.591	2.538	2.494	2.352	2.276	2.194	2.151	2.106	2.059	2.01
17	4.451	3.592	3.197	2.965	2.81	2.699	2.614	2.548	2.494	2.45	2.308	2.23	2.148	2.104	2.058	2.011	1.96
18	4.414	3.555	3.16	2.928	2.773	2.661	2.577	2.51	2.456	2.412	2.269	2.191	2.107	2.063	2.017	1.968	1.917
19	4.381	3.522	3.127	2.895	2.74	2.628	2.544	2.477	2.423	2.378	2.234	2.156	2.071	2.026	1.98	1.93	1.878
20	4.351	3.493	3.098	2.866	2.711	2.599	2.514	2.447	2.393	2.348	2.203	2.124	2.039	1.994	1.946	1.896	1.843
30	4.171	3.316	2.922	2.69	2.534	2.421	2.334	2.266	2.211	2.165	2.015	1.932	1.841	1.792	1.74	1.684	1.622
40	4.085	3.232	2.839	2.606	2.45	2.336	2.249	2.18	2.124	2.077	1.925	1.839	1.744	1.693	1.637	1.577	1.509
60	4.(X)l	3.15	2.758	2.525	2.368	2.254	2.167	2.097	2.04	1.993	1.836	1.748	1.649	1.594	1.534	1.467	1.389
120	3.92	3.072	2.68	2.447	2.29	2.175	2.087	2.016	1.959	1.911	1.751	1.659	1.554	1.495	1.429	1.352	1.254
∞		2.996	2.605	2.372	2.214	2.099	2.01	1.938	1.88	1.831	1.666	1.571	1.459	1.394	1.318	1.221	1

ν_1 = Degrees of freedom.
ν_2 = Degrees of freedom.

In the normal distribution and t test for the difference between two means, the difference between the two observed means was calculated and hypothesized to have come from populations with the same means. No statistic has been discovered yet that provides for the difference between two estimated variances. However, R.A. Fisher discovered a distribution of the F statistic given in Equation 13.13 and corresponding to the function $1/2 \log_e [s^2(X_2)/s^2(X_1)]$. Snedecor modified this to give the values corresponding to $s^2(X_2)/s^2(X_1)$ and named the ratio F in honor of Fisher. Therefore,

$$F = \frac{s^2(X_2)}{s^2(X_1)} \tag{13.13}$$

The terms $s^2(X_1)$ and $s^2(X_2)$ are the variances of the variables X_1 and X_2, calculated from n_1 measurements of X_1 and n_2 measurements of X_2. From Equation 13.13, one notes that F values have been tabulated in terms of an assumed level and two degrees of freedom: $v_1 = n_1 - 1$ and $v_2 = n_2 - 1$. This test provides information on whether one variance is larger than another, not whether two variances are significantly different. This effectively corresponds to the one-sided test employed in the comparison of means. To use the tables for a two-sided test, the indicated probability levels must be doubled.

Like the chi-square distribution, the F distribution is of importance in problems of statistical inference. It is used, for example, as follows:

1. To test whether, on the basis of sample data, there is evidence of a significant difference in the variance of two normally distributed populations.
2. To test whether, on the basis of sample data, there are significant differences among the averages of a number of normal populations. (This test is known as the analysis of variance [ANOVA].)[1]
3. To evaluate significance in least-squares regression analyses. In these applications, the parameters v_1 and v_2 are generally referred to as the degrees of freedom of the F distribution.

For example, one would use Tables 13.5 and 13.6 to obtain values for $F_{0.01;5,3}$ and $F_{0.95;3,12}$. From Table 13.2

$$F_{0.01;5,3} = 28.2$$

and from Table 13.2

$$F_{0.95;3,12} = 3.49$$

From Equation 13.11,

$$F_{0.95;3,12} = \frac{1}{F_{0.05;12,3}}$$

$$= \frac{1}{8.74}$$

$$= 0.114$$

Now consider the following case. The variance of the concentration of contaminated soil is known to be 0.417. A second variance is estimated to be 0.609 from 41 samples. Is the estimated variance significantly larger than the known variance of 0.417? Perform the test at both the 0.01 and 0.05 levels. The key statistic is calculated from Equation 13.13. Note once again that the larger variance must be placed in the numerator:

$$F = \frac{s^2(X_2)}{s^2(X_1)}$$

$$= \frac{0.609}{0.417}$$

$$= 1.46$$

The tabulated values are obtained from Tables 13.5 and 13.6 by first noting that $v_1 = \infty$.

1. For the 0.01 level,

$$F_{0.01;40,\infty} = 1.59$$

Because the tabulated value is greater than the calculated statistic, the hypothesis that the two variances are equal is accepted; that is, accept H_0.
2. For the 0.05 level,

$$F_{0.05;40,\infty} = 1.40$$

The hypothesis at this level is rejected.

From the preceding results, one can conclude that there is less than a 5% chance (or 0.05 probability) but more than a 1% chance that the variances came from the same population.

Consider also the following chemical laboratory study. Twenty-five ash weight measurements from an analytical balance produce results for the sample variance of $s_1^2 = 0.45$ mg. Thirty-one weight measurements with another analytical balance produced a sample variance of $s_2^2 = 0.20$ mg. Does the first balance have a σ_1 greater than σ_2? Test this hypothesis at the 1% level of significance.

For this case,

$$H_0 : \sigma_1^2 = \sigma_2^2$$

$$H_1 : \sigma_1^2 > \sigma_2^2$$

This is a one-sided (right-tail) test. Noting that $\alpha = 0.01$, $v_1 = 24$, and $v_2 = 30$, Table 13.5 yields

$$F_{0.01;24,30} = 2.47$$

and

$$\frac{S_1^2}{S_2^2} = \frac{45}{20} = 2.25$$

The critical region for the test is obtained from Table 13.6. Because 2.25 is not greater than 2.47, that is, $2.25 < 2.47$, the hypothesis H_0 is accepted.

OTHER OCCASIONALLY USED CONTINUOUS PROBABILITY DISTRIBUTIONS

The occasionally used continuous distributions are listed in the following.

1. Pascal's distribution
2. Geometric distribution
3. The Cauchy distribution
4. The beta distribution
5. The gamma distribution
6. The uniform (also referred to as a square wave or rectangular) distribution (generally employed when only *two* values of the population are known)
7. The triangular distribution (generally employed with two values of the population involving one higher value)
8. The Maxwell distribution

The reader is referred to the statistics literature if developmental material and illustrative examples are desired.

ILLUSTRATIVE EXAMPLES

Illustrative Example 13.1

Obtain the confidence coefficient t for either a left or right tail of Student's distribution in the following cases:

1. For $v = 10$, tail area $= 0.05$
2. For $v = 10$, tail area $= 0.01$
3. For $v = 20$, tail area $= 0.10$
4. For $v = 20$, tail area $= 0.025$
5. For $v = 100$, tail area $= 0.05$
6. For $v = \infty$, tail area $= 0.05$

Solution
Refer to Table 13.2.

1. With $v = 10$, 95% (0.95 tail), $t_{10} = 1.81$
2. With $v = 10$, 99% (0.99 tail), $t_{10} = 2.76$
3. With $v = 20$, 90% (0.90 tail), $t_{20} = 1.32$
4. With $v = 20$, 97.5% (0.975 tail), $t_{20} = 2.09$

5. With $v=100$, 95% (0.95 tail), t_{100} (interpolated) $=1.66$
6. With $v=\infty$, 95% (0.95 tail), $t_{\infty}=1.645$ (normal distribution)

Note that the last calculated value corresponds exactly to that provided by a normal distribution, that is, 1.645 (see also Chapter 9).

Illustrative Example 13.2

Ozone measurements in parts per million in an air pollution study were obtained from samples of 40 sites selected at random in a large city. The sample mean, $\bar{X}$, was 10.1. The sample standard deviation, s, was 2.3. Obtain a 95% confidence interval for the population mean ozone measurement for this city.

Solution
First, note that the sample size is large. Select a statistic involving the sample mean and population mean and having a known pdf under the assumptions $(\bar{X}-\mu)/(\sigma/\sqrt{n})$ is approximately distributed as a standard normal variable for $n=40$. Construct an inequality concerning the statistic selected, such that the probability that the inequality is true equals the desired confidence coefficient. From the normal table, $P(-1.96 < Z < 1.96) = 0.95$ if Z is a standard normal variable. Therefore,

$$P\left(-1.96 < \frac{\bar{X}-\mu}{\sigma/\sqrt{n}} < 1.96\right) = 0.95$$

Solve the preceding equation for the population mean μ,

$$P\left(-1.96 < \frac{\bar{X}-\mu}{\sigma/\sqrt{n}} < 1.96\right)$$
$$-1.96\left(\sigma/\sqrt{n}\right) < \bar{X}-\mu < 1.96\left(\sigma/\sqrt{n}\right)$$
$$-\bar{X}-1.96\left(\sigma/\sqrt{n}\right) < -\mu < -\bar{X}+1.96\left(\sigma/\sqrt{n}\right)$$

Multiplication by −1 requires reversal of the inequality signs,

$$\bar{X}+1.96\left(\sigma/\sqrt{n}\right) > \mu > \bar{X}-1.96\left(\sigma/\sqrt{n}\right)$$

or

$$\bar{X}-1.96\left(\sigma/\sqrt{n}\right) < \mu < \bar{X}+1.96\left(\sigma/\sqrt{n}\right)$$

Substitute the observed values from the sample to obtain the 95% confidence interval for μ. Note that $\bar{X}=10.1$, $n=40$, σ is unknown and is replaced by s, the sample standard deviation whose value is 2.3:

$$10.1 - 1.96\left(2.3/\sqrt{40}\right) < \mu < 10.1 + 1.96\left(2.3/\sqrt{40}\right)$$
$$9.39 < \mu < 10.81$$

Therefore, $9.39 < \mu < 10.81$ is the 95% confidence interval for μ.

Illustrative Example 13.3

A sample of 82 PCB readings at a Superfund site has a mean contaminant concentration of 650 ppm and a standard deviation of 42 ppm. Obtain the 98% confidence limits for the mean concentration.

Solution

Based on the problem statement,

$$\overline{X} = 650; \quad s = 42; \quad n = 82$$

Assume that $\sigma = s$. Therefore,

$$\frac{\sigma}{\sqrt{n}} = \frac{s}{\sqrt{n}} = \frac{42}{\sqrt{82}}$$
$$= 4.638$$

For a two-tailed (with Z normally distributed),

$$P\left(-2.33 < Z < 2.33\right) = 0.98$$

or

$$P\left(-2.33 < \frac{\overline{X} - \mu}{\sigma/\sqrt{n}} < 2.33\right) = 0.98$$

Solving gives

$$\mu = 650 \pm 10.79$$

or

$$639.21 < \mu < 660.79$$

Illustrative Example 13.4

Calculate the probability that 13 atmospheric pressure readings (mbar) from a population with variance of 0.8 will have a sample variance greater than 1.6.

Solution

Assume the population pressure readings are normally distributed. As before, the random variable

$$\frac{(n-1)s^2}{\sigma^2}$$

has a chi-square distributed with $n - 1$ or degrees of freedom. The describing equation is written as

$$P\left(s^2 > 1.6\right) = \left(\frac{(n-1)s^2}{\sigma^2} > \frac{12(1.6)}{0.8}\right)$$

$$= \left(\chi^2 > 24\right)$$

From Table 13.3 with $v = 12$,

$$P\left(s^2 > 1.6\right) \cong 0.021 \left(\text{linear interpolation}\right)$$

As expected, the probability is low.

Illustrative Example 13.5

Outline a procedure that employs the chi-square distribution in testing goodness of fit.

Solution

One of the principal applications of the chi-square distribution is testing goodness of fit. A test of goodness of fit is a test of the assumption that the pdf of a random variable has some specified form. The chi-square test of goodness of fit utilizes as a test statistic

$$\sum_{i=1}^{k} \frac{(f_i - e_i)^2}{e_i} \tag{13.14}$$

where

  k is the number of categories in which the observations on the random variable
    have been classified
  f_i is the observed frequency of the ith category
  e_i is the corresponding theoretical frequency, $i = 1,..., k$ under the assumption
    being tested

If the assumption being tested is true, then the chi-square test statistic is approximately distributed as a random variable having a chi-square distribution with degrees of freedom $v = k - m - 1$, where m is the number of parameters estimated from the data in order to obtain the theoretical frequencies for each of the k

categories in which the data have been classified. Large values of the chi-square test statistic lead to rejection of the assumption being tested.

The P-value of the test is the probability that a random variable having a chi-square distribution with v degrees of freedom exceeds the calculated value of the chi-square test statistic. When the P-value is small, usually less than 0.05, then the assumption being tested is rejected. This procedure is detailed in several applications in Section IV.

Illustrative Example 13.6

The data in Table 13.7 are provided for two methods of measuring for the PCB concentration at a Superfund site. Do the variances produced by the two methods differ significantly? Employ a 10% level of significance.

Solution

F calculated for the two methods is

$$F = \frac{s_1^2}{s_2^2} = \frac{(260)^2}{(201)^2}; \quad s_1 = 260, \quad s_2 = 201$$

$$= 1.67$$

At the 10% level of significance with $v_1 = 7$ and $v_2 = 11$, one obtains (because this is a two-sided test)

$$F_{0.05;7,11} = 3.01 \,(\text{by interpolation})$$

Because $3.01 > 1.67$, the difference between the variances is *not* significant.

Illustrative Example 13.7

The sample variance of yield (in percentage) from two nuclear reactors is provided in Table 13.8. If

$$H_0 : \sigma_1^2 = \sigma_2^2$$

$$H_1 : \sigma_1^2 > \sigma_2^2 \quad \text{or} \quad H_0 : \sigma_1^2 < \sigma_2^2$$

obtain the critical region at a 10% level of significance.

TABLE 13.7

PCB Data

Method	Sample Size	Standard Deviation (ppm)
1	8	260
2	12	201

TABLE 13.8

Reactor Data

Reactor	Sample Size	Sample Variance (s^2)
1	8	9.9
2	8	3.8

Solution

Note that F is distributed as

$$F_{7,7}$$

The critical region for this two-tailed test is therefore

$$F < F_{0.05;7,7} \quad \text{and} \quad F > F_{0.05;7,7}$$

From Table 13.6,

$$F < 3.79 \quad \text{and} \quad F > 0.264$$

For the given test,

$$F = \frac{9.9}{3.8} = 2.60$$

Because $0.264 < 0.260 < 3.79$, H_0 is accepted. The reader is left the exercise of showing that H_0 would be rejected at the 2% level of significance ($F = 6.99$).

REFERENCES

1. L. Theodore and F. Taylor, *Probability and Statistics*, Theodore Tutorials (originally published by USEPA, RTP, NC), East Williston, NY, 1996.
2. S. Shaefer and L. Theodore, *Probability and Statistics Applications in Environmental Science*, CRC Press/Taylor & Francis Group, Boca Raton, FL, 2007.
3. http://www.statsoft.com/textbook/stable.html.

Section IV

Applications

The bearings of this observations lays in the application in it.

Charles Dickens (1812–1870)

This final section is concerned with environmental risk applications for specific industries. The industries selected, along with their corresponding chapters, are listed as follows:

Chapter 14: Chemicals and Refineries
Chapter 15: Energy and Power
Chapter 16: Manufacturing and Electronics
Chapter 17: Pharmaceuticals
Chapter 18: Military and Terrorism
Chapter 19: Travel/Aerospace/Weather
Chapter 20: Nanotechnology

Each chapter contains solved applications involving many of the probability distributions presented in Sections II and III specific to each industry. The last chapter (Chapter 20) was a last-minute add-on because of the recent interest in this new technology. Finally, several of the examples/problems involve international applications.

14 Chemicals and Refineries

There are 19 illustrative examples in this chapter. They key on environmental risk calculations that involve both discrete and continuous probability distributions. Discrete distributions that receive attention include binomial, multinomial, hypergeometric, and Poisson; continuous distributions include: normal, log-normal, exponential, Weibull, Monte Carlo, chi-square (χ^2), and F distributions. The presentation of the problems attempts to follow this order, that is, as they were reviewed in Sections II and III. Two of the applications are based on activities at the international level.

Please note that many of the applications were adapted from the literature.[1–3]

APPLICATION 14.1

Pumps are devices with glands that leak. Packed glands are cheap and simple and rarely fail catastrophically, but there is usually a slight leak, and they need regular adjustment. Mechanical seals are therefore widely used, often with a second seal to contain any leakage. They require skilled assembly and accurate shaft alignment. Canned motors and magnetic drives eliminate the need for seals, but their initial cost is high. When experience shows that pumps handling problematic liquids are particularly liable to leak, for example, those handling very hot or cold liquids, remotely operated *emergency isolation valves* should be fitted in the suction lines so that leaks can be stopped from a safe distance.[2] All glands may fail catastrophically if there is a bearing failure. If the quantity of liquid in the suction vessel is large, then emergency isolation valves should be installed even though the pumps have no history of failure. Otherwise, if failure should occur, a large quantity of liquid will be discharged.

If a pumping system in the environmental control section of a refinery must have a reliability of 99.993%, how many pumps are required in a parallel system if each pump has a reliability of 94.2% due to pump leakage problems? Assume the reliabilities of the pumps are equal.

Solution

Refer to Equation 11.8 (in Chapter 11). The required reliability for this parallel system is

$$R_p = 1 - (1 - R_1)(1 - R_2)\ldots(1 - R_i)\ldots(1 - R_n) \tag{11.8}$$

where R_i is the fractional reliability of pump i.

Since the reliabilities of the pumps are assumed equal,

$$R_1 = R_2 = \cdots = R_n = R$$

then

$$R_p = 1 - (1 - R)^n$$

Noting that

$$R_p = 0.99993$$

and

$$R = 0.942$$

the solution to the equation yields $n = 3.4$. Therefore,

$$n = 4$$

APPLICATION 14.2

A random variable X denoting the useful life in decades of a distillation column processing liquid waste in Southeast Asia has the probability distribution function (pdf)

$$f(x) = \left(\frac{3}{8}\right)x^2; \quad 0 < x < 2$$

$$= 0; \quad \text{elsewhere}$$

Determine the cumulative distribution function (cdf) of X.

Solution

The cdf of X is given by

$$f(x) = P(X \le x)$$

$$= \int_{-\infty}^{\infty} f(x) dx$$

Therefore,

$$F(x) = 0; \quad x \le 0$$

$$F(x) = \int [(3/8)x^2] dx$$

$$= \frac{x^2}{8}; \quad 0 < x < 2$$

$$F(x) = 1; \quad x \ge 2$$

APPLICATION 14.3

Road tankers (tank trucks) and rail tankers (tank cars) are widely used for the conveyance of liquids to and from chemical plants and refineries, and their record in the United States is remarkably good. Most accidents involving tankers occur while they are being filled or emptied, and the following are typical:

1. A tanker is overfilled or it is emptied without a vent, such as a manhole.
2. A leak occurs because the wrong type of *hose* is used, or a damaged hose, or the hose is not correctly fitted.
3. A tanker is moved before the hose is disconnected.
4. A tanker is splash filled with a flammable liquid, and the vapor is ignited by a discharge of *static electricity*. Sometimes a tanker is splash filled with a high-boiling liquid such as paraffin, and vapor in the tanker from the previous load is ignited (switch filling).
5. The wrong liquid is put into a tanker or it is emptied into the wrong tank.

The time, X, to failure in months of a tanker in a chemical facility has a pdf specified as follows:

$$f(x) = (1/10)e^{-x/10}; \quad x > 0$$

Calculate the expected value of X and of X^2. Refer to Equation 4.22.

Solution

The expected value of X is

$$E(X) = \int_0^\infty x(1/10)e^{-x/10}dx = 10 \ (mo)$$

The expected value of X^2 is

$$E(X^2) = \int_0^\infty x^2(1/10)e^{-x/10}dx = 200 \ (mo)^2$$

The calculation of the variance is left as an exercise for the reader.

APPLICATION 14.4

The following scenario is under study at a chemical plant processing hazardous waste. Suppose that the probability that a pump switch fails on any throw is 0.005 and that successive throws are independent with respect to failure. If the switch fails for the first time on the throw x, it must have been successful on each of the preceding $x - 1$ trials. What is the pdf of X given the earlier conditions?

Solution

This is a relatively easy application to solve. The pdf of X is given by

$$f(x) = (0.995)e^{x-1}(0.005); \quad x = 1,2,3,\ldots,n$$

See also Chapter 4 for a similar application.

APPLICATION 14.5

There have been a host of companion terms used to describe chemical reactors; among these are *reaction unit, reactor, oxidizer, afterburner, reactor device, burner,* and *organic (or inorganic) reactor.* Whichever term is used, the overall process of chemical reactions is best characterized by phenomena occurring in a *chemical reactor.* In effect, an incinerator is one of a number of units that fits into the class of what industry describes as a *chemical reactor.* With this in mind, one may apply chemical reactor principles to either design and/or predict the performance of these units in this broad category. In any event, this unit is the equipment in which chemical reactions take place.

From a physical perspective, chemical reactors are usually in the form of

1. A cyclinder or cylindrical vessel or tank in which the entire reaction mixture is contained in one part or the entire space
2. A spherical tank, in which the entire reaction mixture is contained in one part or the entire space
3. A tube, in which the reacting mixture flows from one end to the other with little or no mixing

The differences between these three kinds of reactors are often described in terms of the degree of mixing. And there are three major classifications of reactors: batch, tank flow, and tubular flow. These three classes of reactors are discussed in the paragraphs that follows.[4–6]

A *batch reactor* is a vessel or container that may be open or closed. Reactants are usually added to the reactor simultaneously. The contents are then mixed (if necessary) to ensure no variations in the concentrations of the species present. The reaction then proceeds. During this period, there is no transfer of mass into or out of the reactor. Because the concentrations of reactants and products change with time, this is a *transient* or *unsteady-state* operation. The reaction is terminated when the desired chemical change has been achieved. The contents are then discharged and often sent elsewhere for further processing.

Another reactor where mixing is important is the *tank flow* or *continuously stirred tank reactor.* This type of reactor also consists of a tank or kettle equipped with an agitator. It may be operated under steady-state or transient conditions. Reactants are fed continuously, and the products are withdrawn continuously. The reactants and products may be in the liquid, gas, or solid state, or a combination of these. If the contents are perfectly mixed, the reactor design problem is greatly simplified for steady-state conditions because the mixing results in uniform concentration, temperature, and so on, throughout the reactor.

The last and most important class of reactor to be observed is the *tubular flow reactor*. The most common type is the single-pass cylindrical tube; another important type is one that consists of a number of tubes in parallel. The reactor(s) may be vertical or horizontal. The feed is charged continuously at the inlet of the tube, and the products are continuously removed at the outlet. If heat exchange with the surroundings is required, the reactor tube is *jacketed*. If adiabatic conditions are required, the reactor is *covered* with insulation. Tubular flow reactors are usually operated under steady-state conditions so that physical and chemical properties do not vary with time. Unlike the batch and tank flow reactors, there is no mechanical mixing.

Successive failures of a chemical reactor's cooling system in months is given by

$$f(x) = 0.005e^{-0.005x}; \quad x > 0$$

$$f(x) = 0; \quad \text{elsewhere}$$

What is the probability that the time in months between successive cooling failures is greater than 2 but less than 5? Interest in this cooling system has surfaced recently because of the potential for a reactor explosion.

Solution

The probability that the time in months between successive cooling failures is greater than 2 but less than 5 is

$$P(2 < X < 5) = \int_{2}^{5} 0.005e^{-0.005x} \, dx$$

$$= 0.015 = 1.5\%$$

APPLICATION 14.6

Compressors, unlike fans and pumps, find only limited use for specialized applications. They are employed primarily to increase the pressure of a fluid in some types of systems, such as when liquids are broken up into tiny droplets (atomized) before entering a unit. This can be accomplished by using a high-pressure stream of air or steam that impinges on the liquid steam and atomizes it. Pressurization of the air or steam is accomplished through the use of compressors. Compressors operate much the same way as pumps and have the same classification category: rotary, reciprocating, and centrifugal.[5-7]

A compressor has an average time to failure of approximately 450 weeks. Assuming the compressor failure distribution can be reasonably described by an exponential equation, estimate the reliability that a compressor will survive 260 weeks.

Solution

For this calculation, the average time to failure, t_f, is

$$t_f = 450 \text{ weeks}$$

For an exponential model (refer to Chapter 11),

$$R = e^{-\lambda t}$$

where λ is the reciprocal of the average time to failure. Thus,

$$\lambda = \frac{1}{t_f} = \frac{1}{450} = 0.00222 \, (\text{weeks})^{-1}$$

The reliability for 260 weeks is therefore

$$R = e^{-0.00222(260)}$$

$$= e^{-0.578}$$

$$= 0.561 = 56\%$$

APPLICATION 14.7

The fractional probability that a crude oil delivery to a refinery will contain sulfur above the regulatory requirement is 0.25. If there were 6 deliveries last month, find the probability that the sulfur content will be above the required limit.

1. Exactly 2 times
2. More than 4 times

Solution

This involves a binomial distribution calculation with

$$n = 6$$

$$p = 0.25$$

$$q = 0.75$$

1. For this case,

$$P(X = 2) = \frac{6!}{(2!)(4!)}(0.25)^2(0.75)^4$$

$$= 0.297 \approx 30\%$$

2. For more than 4 excursions, the limit has to be exceeded 5 or 6 times. Thus,

$$P(X > 4) = P(X = 5) + P(X = 6)$$

$$P(X = 5) = \frac{6!}{(1!)(5!)}(0.25)^5(0.75)^1$$

$$= 0.00440 = 0.440\%$$

$$P(X = 6) = \frac{6!}{(0!)(6!)}(0.25)^6$$

$$= 0.000244 = 0.0244\%$$

Therefore,

$$P(X > 4) = 0.00464 = 0.464\%$$

APPLICATION 14.8

A cargo ship containing crude oil arrives at a refinery port once a month. The oil may be contaminated with excessive amounts of sulfur (S), mercury (M), ash (A), cadmium (C), sodium (Na), and chlorine (Cl). If the probability of oil being contaminated with any of these is one-sixth, find the probability P of obtaining S and M being exceeded exactly twice and the other four contaminants exactly once in the next eight shipments.

Solution

The multinomial distribution is applied to obtain this answer.

$$P(2,2,1,1,1,1) = \frac{8!}{(2!)(2!)(1!)(1!)(1!)(1!)}\left(\frac{1}{6}\right)^2\left(\frac{1}{6}\right)^2\left(\frac{1}{6}\right)\left(\frac{1}{6}\right)\left(\frac{1}{6}\right)\left(\frac{1}{6}\right)$$

$$= \frac{35}{5832}$$

$$= 0.006 = 0.6\%$$

APPLICATION 14.9

The parts-per-million concentration of a particular toxic in a refinery discharge stream is known to be normally distributed with mean $\mu = 400$ and standard deviation $\sigma = 8$. Calculate the probability that the toxic concentration, C, is between 392 and 416 ppm.

Solution

As noted in Chapter 9, if C is normally distributed with a mean μ and a standard deviation σ, then the random variable $(C - \mu)/\sigma$ is also normally distributed with mean 0 and standard deviation 1 with the term $(C - \mu)/\sigma$ referred to as a *standard normal variable,* and its pdf is called a *standard normal curve.* The probabilities about a standard normal variable Z can be determined from the tables in Chapter 9. Since C is normally distributed with a mean $\mu = 400$ and a standard deviation $\sigma = 8$, then $(C - 400)/8$ is a standard normal variable and

$$P\left(392 < C < 416\right) = P\left\{\frac{392 - 400}{8} < \left[\frac{(C - 400)}{8}\right] < \frac{416 - 400}{8}\right\}$$

$$= P\left\{-1 < \left[\frac{(C - 400)}{8}\right] < +2\right\} = P(-1 < Z < +2)$$

From tabulated or graphical values (see Table 9.1),

$$P\left(392 < C < 416\right) = 0.341 + 0.477$$

$$= 0.818 = 81.8\%$$

APPLICATION 14.10

The dioxin concentration measurements ($\mu g/m^3$) in a refinery fluid bed slip stream is approximately normally distributed with a mean $\mu = 102$ and a standard deviation $\sigma = 3.75$. Find the percentage of readings that are

1. Between 99 and 106.5 (both in $\mu g/m^3$)
2. At least 108 $\mu g/m^3$

Solution

1. Once again, transform into standard units:

$$Z_1 = \frac{99 - 102}{3.75} = -0.80$$

$$Z_2 = \frac{106.5 - 102}{3.75} = 1.20$$

For this case, $Z_1 < Z < Z_2$. Therefore,

$$P(99 \le X \le 106.5) = P(-0.8 \le Z \le 1.2)$$

$$= 0.3849 + 0.2881$$

$$= 0.6730 = 67.3\%$$

In effect, approximately two-third of the readings will be between 99 and 106.5 $\mu g/m^3$.

2. Transform into standard units:

$$Z_1 = \frac{108 - 102}{3.75} = 1.6$$

Therefore,

$$P(X \ge 108) = P(Z \ge 1.6)$$

$$= 0.5 - 0.4452$$

$$= 0.0548 = 5.5\%$$

APPLICATION 14.11

Frictional losses in pipes can take several forms. An important engineering problem is the calculation of these losses. It can be shown that the fluid can flow in either of two modes—laminar or turbulent. For laminar flow, an equation is available from basic theory to calculate friction loss in a pipe. In practice, however, fluids (particularly gases) are rarely moving in laminar flow.

Since the two methods of flow are so widely different, a different equation describing frictional resistance is to be expected in the case of turbulent flow from that which applies in the case of laminar flow. However, it can be shown that both cases may be handled by one relationship in such a way that it is not necessary to make a preliminary calculation to determine whether the flow is taking place above the critical Reynolds number or below it.[5–7]

One can theoretically derive the pressure drop term ΔP for laminar flow.[5–7] The equation takes the form

$$\Delta P = h_f = \frac{32 \mu v L}{\rho g_c D^2} \tag{14.1}$$

for a fluid flowing through a straight cylinder of diameter D and length L. A friction factor, f, that is dimensionless may now be defined as (for laminar flow)

$$f = \frac{16}{Re} \tag{14.2}$$

so that Equation 14.1 takes the form

$$h_f = \Delta P = \frac{4 f L v^2}{2 g_c D} \tag{14.3}$$

Although the earlier equation describes friction loss or the pressure drop across a conduit of length L, it can also be used to provide the pressure drop due to friction per unit length of conduit, for example, $\Delta P/L$ by simply dividing the earlier equation by L.

The pressure loss due to friction in the flow of a carcinogenic at a chemical plant through a circular pipe of known length is known to be a function of the carcinogenic chemical's velocity in the following form:

$$\Delta P = av^b \tag{14.4}$$

where
ΔP is the pressure loss, psi
v is the average velocity of the water, ft/s
a,b is the empirical constants

Evaluate a and b for the flow of the chemical at 70°F through a smooth pipe with a 1 inch inside diameter which is 100 ft long using the experimental data below in Table 14.1.

Solution

Rewrite Equation 14.4 in the following form:

$$\log(\Delta P) = b \log v + \log a \tag{14.5}$$

This equation will give a straight line of slope b and intercept a on log–log coordinates. The slope and intercept are obtained from a log ΔP – log v plot.

$$b = \text{slope} = 1.76$$

$$a = \text{intercept} = \Delta P \text{ at } v = 1$$

$$= 0.26$$

TABLE 14.1

Pressure Drop—Velocity Data

v, ft/s	ΔP (psi)
1	0.26
3	1.75
5	4.5
7	8.2
9	12.5
12	21.0
15	32.5

Therefore, the describing equation is

$$\Delta P = 0.26 v^{1.76}$$

APPLICATION 14.12

The probability that a fan in a refinery will not survive for more than 5 years is 0.90. How often should the fan be replaced? Assume that the time to failure is exponentially distributed and that the replacement time should be based on the fan's expected life.

Solution

This requires the calculation of μ in the exponential model with units of $(years)^{-1}$. Once again (see also Chapter 11),

$$F(t) = 1 - e^{-\lambda t}$$

Based on the information provided,

$$P(T \le 5) = 0.90$$

or

$$0.90 = 1 - e^{-\lambda(5)}$$

Solving for λ gives

$$\lambda = -\frac{1}{5} \ln(1 - 0.90)$$
$$= 0.4605$$

Because the expected time (or life), $E(T)$, is

$$E(T) = \frac{1}{\lambda}$$
$$= \frac{1}{0.4605}$$
$$= 2.17 \text{ years}$$

Therefore, the fans should be replaced in approximately 2 years.

APPLICATION 14.13

An electronic control system in a refinery consists of three components (A, B, and C) connected in parallel. If the time to failure for each component is exponentially distributed and mean times of failures for components A, B, and C are 600, 300, and 200 weeks, respectively, determine the probability the control system will fail before 365 weeks.

Solution

For this series system, the probability of failure is (see Chapter 11)

$$P(F) = P(\text{all components fail})$$

$$= (1 - P_A)(1 - P_B)(1 - P_C)$$

where P_1 is the probability of surviving 365 weeks. The system reliability is

$$R = 1 - P(F)$$

$$= 1 - (1 - P_A)(1 - P_B)(1 - P_C)$$

Since the time to failure for each component is exponentially distributed,

$$P_A = e^{-\left(\frac{365}{600}\right)} = 0.544$$

$$P_B = e^{-\left(\frac{365}{300}\right)} = 0.296$$

$$P_C = e^{-\left(\frac{365}{200}\right)} = 0.161$$

Therefore, the system reliability is

$$R = 1 - (1 - 0.544)(1 - 0.296)(1 - 0.161)$$

$$= 0.731 = 73.1\%$$

The probability the system will fail within that same period is simply

$$P = 1 - R(\text{fractional basis})$$

$$P = 1 - 0.731$$

$$= 0.269 = 27\%$$

Refer to Illustrative Example 11.2 (Chapter 11) for a similar example employing the above data.

APPLICATION 14.14

Tubing and other conduits are used for the transportation of gases, liquids, and slurries. These ducts are often connected and may also contain various valves and fittings, including expansion and contraction joints. There are four basic types of connecting conduits:

1. Threaded
2. Bell-and-spigot
3. Flanged
4. Welded

Extensive information on these connection classes is available in the literature.[5–7]

Because of the diversity of types of systems, fluids, and the environments in which valves must operate, a vast array of valve types have been developed. Examples of the common types are the globe valve, gate valve, ball valve, plug valve, pinch valve, butterfly valve, and check valve. Each type of valve has been designed to meet specific needs. Some valves are capable of throttling flow, other valve types can only stop flow, others work well in corrosive systems, and others handle high-pressure fluids.[5–7]

Valves have two main functions in a pipeline: to control the amount of flow or to restrict the flow completely. Of the many different types of valves employed in practice, the most commonly used are the gate valve and the globe valve. The *gate valve* contains a disk that slides at right angles to the flow direction. This type of valve is used primarily for on/off control of a liquid flow. Because small lateral adjustments of the disk can cause significant changes in the flow cross-sectional area, this type of valve is not suitable for accurate adjustment of flow rates. As the fluid passes through the gate valve, only a small amount of turbulence is generated; the direction of flow is not significantly altered, and the flow cross-sectional area inside the valve is only slightly less than that of the pipe. As a result, the valve causes only a minor pressure drop. Problems with abrasion and erosion for the disk arise when the valve is used in positions other than fully open or fully closed. Unlike the gate valve, the *globe* valve—so called because of the spherical shape of the valve body—is designed for more sensitive flow control. In this type of valve, the liquid passes through the valve in a somewhat tortuous or circuitous route. In one form, the seal is a horizontal ring into which a plug with a slightly beveled edge is inserted when the stem is closed. Good control of flow is achieved with this type of valve, but at the expense of a higher pressure loss than with a gate valve.[5–7]

Consider a valve at a chemical facility's water treatment plant whose time to failure *T*, in hours, has a Weibull pdf with parameters $\alpha = 0.01$ and $\beta = 0.50$. This results in

$$f(t) = (0.01)(0.50)t^{0.5-1}e^{-(0.01)t^{0.5}}$$

as the Weibull pdf of the failure rate of the component under consideration over its entire time domain. Estimate the probability that the valve will operate more than 8100 days.

Solution

$$P(T < 8100) = \int_0^{8100} f(t)\,dt$$

$$= \int_0^{8100} 0.005t^{-0.5}e^{-(0.01)t^{0.5}}\,dt$$

$$= e^{-(0.01)t^{0.5}}\Big|_0^{8100}$$

$$= -0.41 + 1$$

$$= 0.59 = 59\%$$

APPLICATION 14.15

A sample of 27 ambient particulate readings at an industrial site has a mean concentration of 92 μg/m³ and a standard deviation of 14 μg/m³. Obtain the 98% confidence limits for the mean concentration.

Solution

Based on the problem statement,

$$\bar{X} = 92; \quad s = 14; \quad n = 27$$

Assume that $\sigma = s$. Therefore,

$$\frac{\sigma}{\sqrt{n}} = \frac{s}{\sqrt{n}} = \frac{14}{\sqrt{27}}$$

$$= 2.69$$

Refer to Chapter 13. For a two-tailed test (with Z normally distributed),

$$P(-2.33 < Z < 2.33) = 0.98$$

or

$$P\left(-2.33 < \frac{\bar{X} - \mu}{\sigma/\sqrt{n}} < 2.33\right) = 0.98$$

Solving gives

$$\mu = 92 \pm 6.27$$

or

$$85.7 < \mu < 98.3$$

APPLICATION 14.16

A Student's t distribution calculation arises in a good number of environmental risk applications. Answer the following two questions:

1. Briefly describe the distribution of means for small samples.
2. Discuss the confidence range or interval as it applies to the t distribution.

Solution

1. The Student's t distribution is employed when dealing with small samples. The graph of the pdf of the Student's distribution is symmetric about 0 for all values of the parameter. The graph is similar to but not as peaked as the standard normal curve. The graph approaches the normal distribution in the limit when the degrees of freedom approach infinity. The term t_r has also been used to designate a random variable having a Student's distribution with ν degrees of freedom. When the sample size is small ($n < 30$), σ is unknown, and the population sample is assumed to be normal, the statistic $(\bar{X} - \mu)/(s\sqrt{n})$ can, as described earlier, be used to generate confidence intervals for the population mean μ.

2. As noted in the earlier discussion of the normal distribution, $\bar{X}$ is an estimate of the true mean μ. The t function provides the distribution of deviations of $\bar{X}$ from μ in terms of probabilities (or relative frequencies). If one rewrites the t equation in terms of plus or minus, the true mean for this class is given by

$$\mu = \bar{X} \pm ts$$

In ordinary language, the true mean can be said, with the tabulated probability of error, to be within the range of the calculated mean included in the limits of plus and minus t times the estimated standard deviation of the mean. It should be once again noted that one is merely applying the confidence interval numbers to a given experiment. It is obviously incorrect to state that the probability is 0.95 (or 0.99, etc.) that the interval contains μ; the latter probability is either 1 or 0, depending on whether μ does or does not lie in this interval. What it does mean is that μ will lie within the interval 95 out of 100 times. It is only when the random interval $\bar{X} \pm ts$ is considered that one can make correct probability statements of the type desired.

APPLICATION 14.17

As part of an environmental risk assessment study, the American Petroleum Institute (API) obtained the following lifetimes (in years) of 15 refineries:

$$37 \ 39 \ 42 \ 45 \ 36 \ 32 \ 36 \ 36 \ 38 \ 43 \ 40 \ 43 \ 37 \ 30 \ 38$$

Assuming the lifetimes are normally distributed, test API's hypothesis H_0: $\sigma = 3$ years against the alternative hypothesis H_1: $\sigma > 3$ years. Use 0.01 as the tolerated probability of a type 1 error.

Solution

The one-tail critical regional statistic is

$$\frac{(n-1)s^2}{\sigma_0^2} > \chi_{0.05}^2$$

For H_0: $\sigma = 3$ the value of σ_0 is 3. Therefore, from Table 13.3 of the chi-square distribution, $\chi_{0.01}^2 = 29.1$ with $v = 14$ (since $v = n - 1$). Therefore, the critical region is specified by

$$\frac{(n-1)s^2}{9} > 29.1$$

The value of the test statistic $(n-1)s^2/\sigma_0^2$ may now be calculated. One can simply show from Equation 3.39 (from Chapter 3) that $s = 4.19$. Therefore, the value of the test statistic is

$$\frac{(14)(4.19)^2}{0} = 27.3$$

The required hypothesis test may now be completed. Accept H_0: $\sigma = 3$ since the observed value of the test statistic, 27.3, does not exceed the critical region value of 23.7. Thus, the data do not support API's conclusion that the standard deviation of refinery lifetimes is 3 years. The standard deviation exceeds 3 years.

APPLICATION 14.18

The data in Table 14.2 are provided for two methods of measuring for the ambient particulate concentration at a residential site in India. Do the variances produced by the two methods differ significantly? Employ a 10% level of significance.

Solution

The application involves application of the F distribution. Refer to Chapter 13. F is calculated for the two methods as

---

TABLE 14.2
Ambient Particulate Concentrations, µg/m³

Method	Sample Size	Standard Deviation (ppm)
1	10	53
2	15	46

$$F = \frac{s_1^2}{s_2^2} = \frac{(53)^2}{(46)^2}; \quad s_1 = 53, \; s_2 = 46$$

$$= 1.33$$

At the 10% level of significance with $v_1 = 9$ and $v_2 = 14$, one obtains (since this is a *two*-sided test) from Table 13.5

$$F_{0.05;9,14} = 2.65 \text{ (by interpolation)}$$

Because 2.65 > 1.67, the difference between the variances is *not* significant.

APPLICATION 14.19

According to state regulations, three thermometers (A, B, C) must be positioned near the outlet afterburner at a chemical plant treating a gas stream containing hazardous air pollution. Assume that the individual thermometer component lifetimes are normally distributed with means and standard deviations given in Table 14.3.

Using the 10 random numbers given in Table 14.4 for each thermometer, simulate the lifetime (time to thermometer failure) of the temperature recording system and estimate its mean and standard deviation, and the estimated time to failure for this system. The lifetime is defined as the time (in weeks) for one of the thermometers to *fail*.

Solution

Let T_A, T_B, and T_C denote the lifetimes of thermometer components A, B, and C, respectively. Let T_S denote the lifetime of the system. The random number

---

TABLE 14.3
Thermometer Data

Thermometer	A	B	C
Mean (weeks)	100	90	80
Standard deviation (weeks)	30	20	10

TABLE 14.4

Thermometer Random Numbers

For A		For B		For C	
0.52	0.01	0.77	0.67	0.14	0.90
0.80	0.50	0.54	0.31	0.39	0.28
0.45	0.29	0.96	0.34	0.06	0.51
0.68	0.34	0.02	0.00	0.86	0.56
0.59	0.46	0.73	0.48	0.87	0.82

generated in Table 14.4 may be viewed as the cumulative probability, where the cumulative probability is the area under the standard normal distribution curve. Because the standard normal distribution curve is symmetrical, the negative values of Z and the corresponding area are once again found by symmetry. For example, as described earlier,

$$P(Z < -1.54) = 0.062$$

$$P(Z > 1.54) = 0.062$$

$$P(0 < Z < 1.54) = 0.5 - P(Z > 1.54)$$

$$= 0.5 - 0.062$$

$$= 0.438$$

The lifetime or time to failure of each component, T is calculated using the equation

$$T = \mu + \sigma Z$$

where
    μ is the mean
    σ is the standard deviation
    Z is the standard normal variable

First determine the values of the standard normal variable, Z, for component A using the 10 random numbers given in the problem statement and the standard normal table. Then, calculate the lifetime of thermometer component A, T_A, using the equation above for T (see Table 14.5).

Next, determine the values of the standard normal variable and the lifetime of the thermometer component for component B (see Table 14.6).

Also, determine the values of the standard normal variable and the lifetime of the thermometer component for component C (see Table 14.7).

TABLE 14.5
Lifetime of Thermometer A (T_A)

Random Number	Z (from Standard Normal Table)	$T_A = 100 + 30\ Z$
0.52	0.05	102
0.80	0.84	125
0.45	–0.13	96
0.68	0.47	114
0.59	0.23	107
0.01	–2.33	30
0.50	0.00	100
0.29	–0.55	84
0.34	–0.41	88
0.46	0.10	97

TABLE 14.6
Lifetime of Thermometer B (T_B)

Random Number	Z (from Standard Normal Table)	$T_B = 90 + 20\ Z$
0.77	0.74	105
0.54	0.10	92
0.96	1.75	125
0.02	–2.05	49
0.73	0.61	102
0.67	0.44	99
0.31	–0.50	80
0.34	–0.41	82
0.00	–3.90	12
0.48	–0.05	89

For each random value of each component, determine the system lifetime, T_S. Because this is a series system, the system lifetime is limited by the component with the minimum lifetime (see Table 14.8).

The mean value, μ, of T_S is therefore

$$\mu = \frac{635}{10} = 63.5 \text{ years}$$

TABLE 14.7
Lifetime of Thermometer C (T_C)

Random Number	Z (from Standard Normal Table)	$T_C = 80 + 10\,Z$
0.14	−1.08	69
0.39	−0.28	77
0.06	−1.56	64
0.86	1.08	91
0.87	1.13	91
0.90	1.28	93
0.28	−0.58	74
0.51	0.03	80
0.56	0.15	81
0.82	0.92	89

TABLE 14.8
Thermometer System Lifetime

T_A	T_B	T_C	T_S
102	105	69	69
125	92	77	77
96	125	64	64
114	49	91	49
107	102	91	91
30	99	93	30
100	80	74	74
84	82	80	80
88	12	81	12
97	89	89	89
Total			635

Calculate the standard deviation, σ, of T_S using the equation

$$\sigma^2 = \frac{1}{n} \sum (T_S - \mu)^2$$

where n is 10, the number in the population. Note that this is not a sample, so that a modified form of the equation applies for σ^2 (see Table 14.9). Therefore,

$$\sigma = \left(\frac{5987}{10} \right)^{0.5} = 24.5 \text{ years}$$

TABLE 14.9

Thermometer Standard Deviation Calculations

System Lifetime (T_S)	$(T_S, \mu)^2$
69	30.25
77	182.25
64	0.25
49	210.25
91	756.25
30	1122.25
74	110.25
80	272.25
12	2652.25
89	650.25
Total	5987

REFERENCES

1. L. Theodore and F. Taylor, *Probability and Statistics*, Theodore Tutorials (originally published by USEPA, RTP, NC), East Williston, NY, 1996.
2. S. Shaefer and L. Theodore, *Probability and Statistics Applications in Environmental Science*, CRC Press/Taylor & Francis Group, Boca Raton, FL, 2007.
3. L. Theodore and R. Dupont, *Environmental Health and Hazard Risk Assessment: Principles and Calculations*, CRC Press/Taylor & Francis Group, Boca Raton, FL, 2013.
4. L. Theodore, *Chemical Reactor Design and Applications for the Practicing Engineer*, John Wiley & Sons, Hoboken, NJ, 2012.
5. L. Theodore, *Chemical Engineering: The Essential Reference*, McGraw-Hill, New York, NY, 2014.
6. J. Reynolds, J. Jeris, and L. Theodore, *Handbook of Chemical and Environmental Engineering Calculations*, John Wiley & Sons, Hoboken, NJ, 2004.
7. P. Abulencia and L. Theodore, *Fluid Flow for the Practicing Chemical Engineer*, John Wiley & Sons, Hoboken, NJ, 2010.

15 Energy and Power

There are 17 applications of an illustrative example nature in this chapter. They key on environmental risk calculations that involve both discrete and continuous probability distributions. Discrete distributions that receive attention include binomial, multinomial, hypergeometric, and Poisson. Continuous distributions include normal, log-normal, exponential, Weibull, chi-square (χ^2), and Monte Carlo. One of the applications is based on activities at the international level.

The bulk of material in this chapter has been adapted from the literature.[1-3]

APPLICATION 15.1

Large low-pressure utility storage tanks are among the most fragile items of plant equipment in use. They are usually designed to withstand a gauge pressure of only 8 in. of water (0.3 psi)—they will burst at about three times this pressure—and if a vacuum is exceeded by more than a small amount. It is not surprising, therefore, that tanks are often damaged.

A random variable X denoting the useful life in years of a storage tank handling explosive chemicals has the probability distribution function (pdf)

$$f(x) = \frac{x^3}{20.25}; \quad 0 < x < 3$$

$$= 0; \quad \text{elsewhere}$$

Find the cumulative distribution function (cdf) of X.

Solution

As before, the cdf of X is given by

$$f(x) = P(X \leq x)$$

$$= \int_{-\infty}^{\infty} f(x)\,dx$$

Therefore,

$$F(x) = 0; \quad x \leq 0$$

$$F(x) = \int \frac{x^3}{20.25}\,dx = \frac{x^4}{81}; \quad 0 < x < 3$$

$$F(x) = 1; \quad x \geq 3$$

APPLICATION 15.2

The oil industry, including the production, transportation, distribution, and consumption of petroleum products, is regulated to minimize the negative effects on the environment.[4] Yet, environmentalists continue to virtually object to all of the activities of the oil industry. No discussion on the environment can avoid the mention of both the *Exxon Valdez* oil spill of 1987 and the 2010 BP explosion in the Gulf of Mexico.[5] While some spills may have involved more oil, no oil spill received as much attention as the BP fiasco and had as much impact on the local ecosystem. Fortunately, the combination of the cleanup efforts and nature appears to have prevented some of the worst predictions of unrecoverable damage. Theodore and Dupont[1] have provided information and illustrative examples on the *Valdez* and BP fiascos. Although it is too early to draw any conclusions regarding the environmental implications associated with BP's failed oil rig, many have described the incident as the worst environmental disaster in U.S. history.

Let X denote the number of oil cargo ships that arrive at a port on a given day. The pdf for X is given by

$$f(x) = \frac{x}{55}; \quad x = 1,2,3,4,5,6,7,8,9,10$$

1. Verify that $f(x)$ is a valid pdf.
2. Calculate the probability that at least 3 ships but less than 6 ships will arrive on a given day.

Solution

1. For $f(x)$ to be valid, the following two conditions must be satisfied:

$$0 \le f(x) \le 1; \quad x \text{ between 1 and 10}$$

and

$$\sum_{1}^{10} f(x) = 1$$

Substituting into the second condition yields

$$\sum_{1}^{10} f(x) = \sum_{1}^{10} \frac{x}{55}$$

$$= \frac{1}{55} + \frac{2}{55} + \frac{3}{55} + \frac{4}{55} + \frac{5}{55} + \frac{6}{55} + \frac{7}{55} + \frac{8}{55} + \frac{9}{55} + \frac{10}{55}$$

$$= \frac{55}{55} = 1.0$$

Thus, both conditions are satisfied, and $f(x)$ is a valid pdf.

2. The probability that at least 3 ships but no more than 6 ships will arrive on a given day can be determined by

$$P(3 \le X < 6) = f(X = 3) + f(X = 4) + f(X = 5)$$

$$= \frac{3}{55} + \frac{4}{55} + \frac{5}{55}$$

$$= \frac{12}{55}$$

$$= 0.218 = 21.8\%$$

APPLICATION 15.3

Determine the reliability of the electrical system in a nuclear power plant shown in Figure 15.1 using the reliabilities indicated under the various components.

Solution

First identify the components connected in parallel: A and B are connected in parallel; D, E, and F are also connected in parallel. Then, compute the reliability of each subsystem of the components connected in parallel. The reliability of the parallel subsystem consisting of components A and B is

$$R_p = 1 - (1 - 0.7)(1 - 0.7) = 0.91$$

The reliability of the parallel subsystem consisting of components D, E, and F is

$$R_p = 1 - (1 - 0.6)(1 - 0.6)(1 - 0.6) = 0.936$$

Multiply the product of the reliabilities of the parallel subsystems by the product of the reliabilities of the components to which the parallel subsystems are connected in series:

$$R_s = (0.91)(0.9)(0.936)(0.9)(0.9) = 0.621$$

The reliability of the whole system is therefore 0.621.or 62.1

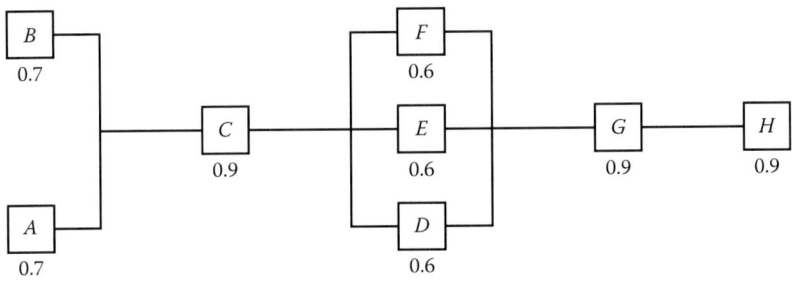

FIGURE 15.1 Diagram of an electrical system.

APPLICATION 15.4

The utility industry principally uses two principal types of heat exchangers:

1. *Double-pipe exchangers*: The simplest type with a concentric pipe arrangement used for cooling and heating; several units can be connected in series to extend their capacity.
2. *Shell and tube exchangers*: The most commonly used for applications in the chemical and allied industries; there are several advantages to this type of heat exchanger: large surface area in a small volume, good mechanical layout, reliance on well-established fabrication techniques, wide range of construction materials available, easily cleaned equipment, and well-established design procedures.

The reader is referred to the literature for the description of other types of heat exchanger units.[6-8]

A procuring agent for a nuclear power plant is asked to sample a lot of 100 heat exchangers. The sample procedure calls for the inspection of 20 exchangers. If there are any bad exchangers, the lot is generally rejected; otherwise it is accepted. The chief engineer has asked the agent the following question. Suppose one bad exchanger is allowed in the sample. What kind of protection would the plant have?

Solution

This application required the application of binomial distribution (see also Chapter 5). One needs to calculate the probability of accepting the lot (with four defective exchangers) if 0 or 1 defective is allowed in the sample. Therefore see also,

$$P(X \leq 1) = P(X = 0) + P(X = 1)$$

$$= 0.442 + \frac{20!}{(1!)(19!)}(0.04)^1 (0.96)^{19}$$

$$= 0.442 + (20)(0.04)(0.4604)$$

$$= 0.442 + 0.368$$

$$= 0.810 = 81.0\%$$

APPLICATION 15.5

Pumps are required to transport liquids, liquid–solid mixtures such as slurries and sludges, auxiliary fuel, and other materials. Pumps are also needed to transport water to or from such peripheral devices as boilers, quenchers, and scrubbers. As indicated earlier, pumps may be classified as reciprocating, rotary, or centrifugal. The reciprocating and rotary types are referred to as *positive-displacement* pumps because, unlike the centrifugal type, the liquid or semiliquid flow is broken into small portions as it passes through the pump.[7-9]

Reciprocating pumps operate by the direct action of a piston on the liquid contained in a cylinder. As the liquid is compressed by the piston, the higher pressure forces it through discharge vales to the pump outlet. As the piston retracts, the next batch of low-pressure liquid is drawn into the cylinder, and the cycle is repeated. The piston may be either directly driven by steam or moved by a rotating crankshaft through a crosshead. The rate of liquid delivery is a function of the volume swept out by the piston and the number of strokes per unit time. A fixed volume is delivered for each stroke, but the actual delivery may be less because of both leakage past the piston and failure to fill the cylinder when the piston retracts. The *volumetric efficiency* of the pump is defined as the ratio of the actual volumetric discharge to the pump displacement volume. For well-maintained pumps, the volumetric efficiency should be at least 95%.[7-9]

As described in Chapter 5, the binomial pdf can be used to calculate the reliability of a redundant system where a redundant system consisting of n identical components is a system that fails only if more than r components fail. Then, X, the number of failures, has the binomial pdf, and the reliability of the redundant system is

$$P(X \leq r) = \sum_{0}^{x} \frac{n!}{x!(n-x)!} p^x q^{n-x} \qquad (15.1)$$

A redundant system at a utility plant consisting of 1 operating pump and 2 on standby can survive 2 pump failures. Assume the pumps are independent with respect to failure and that each has a probability of failure of 0.10. What is the probability of system failure?

Solution

The system consists of three pumps, 1 operating and 2 on standby. Therefore, $n=3$. The system can survive the failure of 2 pumps. Therefore, $r=2$. The probability of a pump failure is 0.10. Therefore, $p=0.10$. Substitute the values of n and p into the binomial pdf (Equation 15.1):

$$f(x) = \sum_{x=0}^{2} \frac{3!}{x!(3-x)!}(0.10)^x (0.90)^{3-x}; \quad x = 0,\ldots,3$$

Sum the binomial pdf from 0 to r to find the reliability.

$$R = \sum_{x=0}^{2} \frac{3!}{x!(3-x)!}(0.10)^x (0.90)^{3-x}$$

$$= 0.999 = 99.9\%$$

The probability of failure, P, is simply given by

$$P = 1 - R$$

Therefore,

$$P = 1 - 0.999$$

$$= 0.001 = 0.1\%$$

APPLICATION 15.6

The probabilities of utilities A, B, and C obtaining a particular type of fuel are equal. Four different categories of each fuel are to be provided. What is the probability that a single utility plant receives all 4 categories of fuel?

Solution

This involves the use of the multinomial distribution. The desired probability is (see Equation 6.9 [in Chapter 6])

$P(1 \text{ utility plant alone getting fuel})$

$= f(4,0,0) + f(0,4,0) + f(0,0,4)$

$= \dfrac{4!}{(4!)(0!)(0!)}(0.333)^4 (0.333)^0 (0.333)^0 + \dfrac{4!}{(0!)(4!)(0!)}(0.333)^0 (0.333)^4 (0.333)^0$

$\quad + \dfrac{4!}{(0!)(0!)(4!)}(0.333)^0 (0.333)^0 (0.333)^4$

$= 0.0123 + 0.0123 + 0.0123$

$= 0.369 = 3.7\%$

Note that for this calculation, each utility plant receives all 4 shipments.

APPLICATION 15.7

Several multipurpose utilities have equal chance to employ fuel in any of the following six forms:

1. Coal
2. Oil
3. Natural gas
4. Nuclear
5. Solar
6. Geothermal

Seven utilities are involved. Find the probability, P, of two employing coal and oil and the others employing only one of the fuels available.

Solution

Employ the multinomial distribution once again.

$$P(2,2,1,1,1) = P = \frac{7!}{(2!)(2!)(1!)(1!)(1!)(1!)} \left(\frac{1}{6}\right)^2 \left(\frac{1}{6}\right)^2 \left(\frac{1}{6}\right) \left(\frac{1}{6}\right) \left(\frac{1}{6}\right)$$

$$= (1260)(0.0278)^2 (0.167)^3$$

$$= 0.00454 = 0.454\%$$

As expected, the probability is low.

APPLICATION 15.8

As noted in Chapter 5, and in an earlier application, a *standby* redundant system is one in which one unit is in the operating mode and n identical units are in the standby mode. Unlike a parallel system where all units in the system are active, in a standby redundancy system, the standby units are inactive. The reliability R of the standby redundancy system described by the Poisson distribution—see also Chapter 8—is given by

$$R = \sum_{x=0}^{n} \frac{e^{-\mu}\mu^x}{x!} \tag{15.2}$$

This is the probability of n or fewer failures in a time period in which the average number of failures is μ.

Consider a standby pumping redundancy system at a coal-fired power plant with 1 operating unit and 2 on standby, that is, a system that can survive 2 failures. If the failure rate is 4 units per year, what is the 6-month reliability of the system?

Solution

The required reliability is a 6-month reliability. Therefore, the associated time period is 6 months. The given failure rate is 4 units per year. Therefore, the associated time period is 1 year or 12 months. Divide the reliability by the time period, that is, 6 divided by 12 yields 1/2. Multiply this result by the failure rate. This is the value of μ.

$$\mu = (1/2)(4) = 2$$

Substitute the value of μ in the Poisson pdf provided in Equation 15.2.

$$f(x) = \frac{e^{-2}2^x}{x!}; \quad x = 0,1,2,\ldots$$

The number of units in standby mode is $n=2$. Therefore, the 6-month reliability is given by

$$R = \sum_{x=0}^{2} \frac{e^{-2}2^x}{x!}$$

$$= e^{-2} + 2e^{-2} + e^{-2}\left(4/2\right)$$

$$= 0.135 + 0.271 + 0.271$$

$$= 0.676 = 67.6\%$$

APPLICATION 15.9

The time to "failure" in weeks of a newly designed power station is normally distributed with a mean and standard deviation of 250 and 5.0 weeks, respectively. Calculate the probability that the new system will fail during a 245- to 260-week period.

Solution

Employing the data provided earlier, generate the standard normal variable Z.

$$Z_1 = \frac{245 - 250}{5.0} = -1.0$$

$$Z_2 = \frac{260 - 250}{5.0} = +2.0$$

Therefore,

$$P\left(245 < Z < 260\right) = P(Z_1 < Z < Z_2)$$

Employing Tables 9.1 and/or 9.2 (in Chapter 9),

$$P = 0.341 + 0.477$$

$$= 0.82 = 82\%$$

APPLICATION 15.10

One of the oldest, simplest, and most efficient methods for removing solid particulate contaminants from gas streams at utilities is by filtration through fabric media. The fabric filter is capable of providing high collection efficiencies for particles as small as 0.1 μm and will remove a substantial quantity of those particles as small as 0.01 μm. In its simplest form, the industrial fabric filter consists of a woven or felted fabric through which dust-laden gases are forced. A combination of factors results in the collection of particles on the fabric filters. When woven fabrics are used, a dust cake eventually forms; this, in turn, acts predominantly as a sieving mechanism. When felted fabrics are used, this dust cake is minimal or almost nonexistent, and the primary filtering mechanisms are a combination of inertial forces, impingement, and so on. These are essentially the same mechanisms that are "applied" to particle collection in wet scrubbers, where the collection media is in the form of liquid droplets rather than solid fibers.[10]

Baghouses may also be characterized and identified according to the method used to remove collected material from the bags. Particle removal can be accomplished in a variety of ways, including shaking the bags, blowing a jet of air on the bags from a reciprocating manifold, or rapidly expanding the bags by a pulse of compressed air. In general, the various types of bag-cleaning methods can be divided into those involving flexing and those involving a reverse flow of clean air.[10]

Let X denote the coded quality of bag fabric used in a particular utility baghouse. Assume that X is normally distributed with a mean 10 and a standard deviation 2. Find k such that $P(X > k) = 0.90$.

Solution

Based on the problem statement,

$$P(X > k) = 0.90$$

$$P\left(\frac{X - \mu}{\sigma} > \frac{k - 10}{2}\right) = 0.90$$

$$P\left(Z > \frac{k - 10}{2}\right) = 0.90$$

For this equation to be valid,

$$\frac{k - 10}{2} = -1.28$$

Solving for k gives

$$k = 7.44$$

APPLICATION 15.11

Particulate discharges from an operation, usually to the atmosphere, consists of a size distribution ranging anywhere from extremely small particles (<1 μm) to very large particles (>100 μm). Particle size distributions are usually represented by a cumulative weight fraction curve in which the fraction of particles less than or greater than a certain size is plotted against the dimension of the particle.

To facilitate recognition of the size distribution, it is useful to plot a size–frequency curve. The size–frequency curve shows the number (or weight) of particles present

TABLE 15.1

Cumulative Distribution Data

Particle Size Range, d_p (μm)	Total (%)	Cumulative %GTSS
<0.62	10	90
0.62–1.0	13	77
1.0–1.2	7	70
1.2–3.0	40	30
3.0–8.0	25	5
8.0–10.0	2	3
>10.0	3	0

for any specified diameter. Because most dusts are comprised of an infinite range of particle sizes, it is first necessary to classify particles according to some consistent pattern. The number or weight of particles may then be defined as that quantity within a specified size range having finite boundaries and typified by some average diameter.

The shape of the curves obtained to describe the particle size distribution generally follows a well-defined form. If the data include a wide range of sizes, it is often better to plot the frequency (i.e., the number of particles of a specified size) against the logarithm of the size. In most cases, an asymmetrical or "skewed" distribution exists; normal probability equations do not apply to this distribution. Fortunately, in most instances, the symmetry can be restored if the logarithms of the sizes are substituted for the sizes. The curve is said to be logarithmic normal (or log-normal) in distribution. Plotting particle diameter versus cumulative percentage therefore generates cumulative distribution plots described in Chapter 10. For log-normal distributions, plots of particle diameter versus either percent less than stated size (% LTSS) or percent greater than stated size (% GTSS) produce straight lines on log-probability coordinates.[7,8,10]

Estimate the mean and the standard deviation from the size distribution information provided in Table 15.1 for a coal-fired utility plant. Use the particle size information for the 15.87% value.

Solution

The use of probability plots is of value when the arithmetic or geometric mean is required because these values may be read directly from the 50% point on a logarithmic probability plot. As noted in Chapter 11, the size corresponding to the 50% point on the probability scale is the *geometric mean diameter*. The geometric standard deviation is given (for % LTSS) by

$$\sigma = \frac{50\% \, \text{size}}{15.87\% \, \text{size}}$$

For % GTSS,

$$\sigma = \frac{50\% \, \text{size}}{15.87\% \, \text{size}}$$

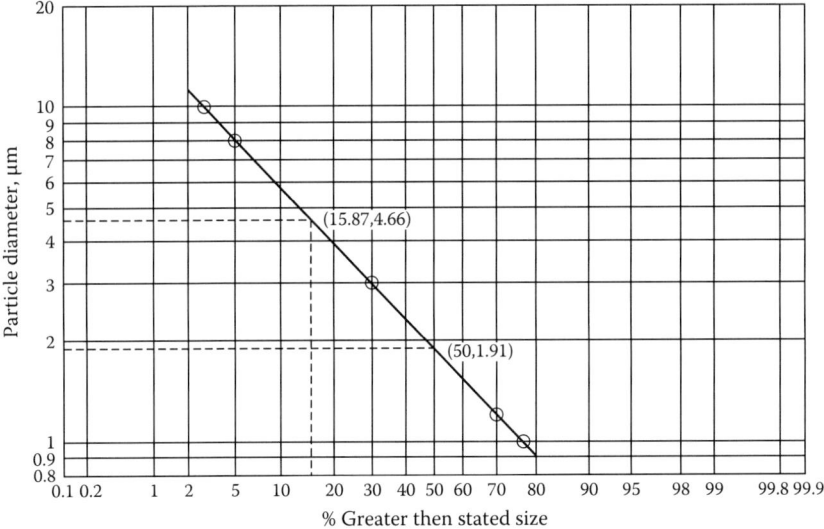

FIGURE 15.2 Cumulative distribution curve for Application 15.11.

The mean, as read from the 50% GTSS point on the graph in Figure 15.2, is approximately 1.9 μm. A value of 1.91 μm is obtained from an expanded plot. From the diagram, the particle size corresponding to the 15.87% point is

$$d_p = (15.87\%) = 4.66 \ \mu m$$

The standard deviation may now be calculated. By definition,

$$\sigma = \frac{d_p(15.87\%)}{d_p(50\%)}$$
$$= \frac{4.66}{1.91}$$
$$= 2.44 \ \mu m$$

The reader may choose to compare this result with that provided in Illustrative Example 10.4.

APPLICATION 15.12

It has been estimated that there are nearly 1 billion tons (nonmetric tons, i.e., 2000 lb each) of proven coal reserves worldwide. This level of availability could last over a century at the current rates of use. Coal is located in almost every country, with recoverable reserves in nearly 75 countries. The largest reserves are in the United

States, Russia, China, and India. After centuries of mineral exploration, the location, size, and characteristic of most countries' coal resources are well known today. What tends to vary even more than the assessed level of the resource, that is, the potentially accessible coal in the ground, is the level classified as proven recoverable reserves. *Proven recoverable reserves* are the tonnage of coal that has been verified by drilling and other means, and are economically and technically extractable. In recent years, there has been a drop in the reserves vs. production (*R/P*) ratio, which has prompted questions over whether the industry has reached a *coal peak*—the point in time at which the maximum global coal production rate is reached and after which the rate enters an *irreversible* decline. However, the recent decline in the *R/P* ratio may be attributed to the lack of incentives to justify increasing reserve numbers.[4]

Where is all the coal in the United States? The two largest producing coal field regions are the Appalachian region, including Pennsylvania, West Virginia, Ohio, western Maryland, eastern Kentucky, Virginia, Tennessee, and Alabama, and the central states region, including Illinois, Indiana, western Kentucky, Iowa, Missouri, Kansas, the Rocky Mountains, and the western states. These coals are mostly sub-bituminous and lignite, which have low sulfur content. Therefore, some of these fields have been developed to meet the increasing demands of electric utilities. The low-sulfur coal permits more economical conformance to environmental regulation, including acid rain legislation.[4]

Assuming an average lifetime of 10 years, calculate the reliability of a coal-fired utility plant in China for

1. 8 years
2. 20 years

Comment on the results.

Solution

For an average lifetime of 10 years,

$$\lambda = \frac{1}{10} = 0.1(\text{year})^{-1}$$

Assume an exponential model applies.

1. For this case, $t = 8$ years, so that

$$R = e^{-(0.1)(8)} = e^{-0.8}$$

$$= 0.50 = 50\%$$

2. For case 2, $t = 1000$ years, so that

$$R = e^{-(0.1)(20)} = e^{-2.0}$$

$$= 0.135 = 13.5\%$$

As expected, as t increases, the reliability of the utilities plant decreases, that is,

$$R(8) = 0.50$$

$$R(20) = 0.135$$

APPLICATION 15.13

The time to failure of the insulation lining in a boiler is exponentially distributed with an expected lifetime of 50 months. Calculate the probability that the lining will survive 20 months. Assume that the replacement time is given by the expected lifetime.

Solution

For an exponential model,

$$F(T) = 1 - e^{-\lambda t} \tag{15.3}$$

For this application,

$$P(T \le 20) = 1 - e^{-\lambda(20)}$$

with

$$\lambda = \frac{1}{50}$$
$$= 0.02$$

Thus,

$$P(T \le 20) = 1 - e^{-\lambda(20)}$$
$$= 1 - e^{-(0.02)(20)}$$
$$= 1 - e^{-0.4}$$
$$= 0.67 = 67\%$$

APPLICATION 15.14

Consider a component in a nuclear power plant whose entire time to failure T, in hours, has a Weibull pdf with parameters $\alpha = 0.01$ and $\beta = 0.50$. This gives

$$f(t) = (0.01)(0.50)t^{0.5-1}e^{-(0.01)t^{0.5}}; \quad t > 0$$

as the Weibull pdf of the failure time of the component under consideration. Estimate the probability that the component will operate less than 10,000 h. See also Chapter 12.

Solution

For the conditions specified,

$$P(T > 10,000) = \int_0^{10,000} f(t)\,dt$$

$$= \int_0^{10,000} 0.005t^{-0.5}e^{-(0.01)t^{0.5}}\,dt$$

$$= e^{-(0.01)t^{0.5}}\Big|_0^{10,000}$$

$$= -1 + 0.368$$

$$= 0.632 = 63.2\%$$

APPLICATION 15.15

The specification for the production of a toxic alloy for a battery employed at a utility plant calls for 23.2% copper. A sample of 10 analyses of the product showed a mean copper content of 23.5% and a standard deviation of 0.24%. Can one conclude at the

1. 0.01 and
2. 0.05

significance levels that the product meets the required specifications.

Solution

Refer to Chapter 13. For this application, H_0: $\mu = 23.2\%$. In addition,

$$t = \frac{\bar{X} - \mu}{\dfrac{s}{\sqrt{n-1}}}$$

$$= \frac{23.5 - 23.2}{\dfrac{0.24}{\sqrt{9}}}$$

$$= 3.75$$

1. At the 0.01 level (see Table 13.2 [in Chapter 13]) with $v = 9$,

$$t_{0.995,9} = 3.25$$

Therefore, reject H_0.

2. At the 0.05 level,

$$t_{0.975,9} = 2.26$$

Therefore, reject H_0 once again.

APPLICATION 15.16

A nuclear facility employs a large number of pumps: 30% are of design A and 70% are of design B. Ten years later, there have been 128 pump failures of design A and 312 of design B. Comment on whether there is evidence of a difference between the two designs.

Solution

If there were no difference between the designs, the expected number of failures would be proportionally the same. For this application,

$$\text{Failures, } A = (0.30)(440)$$

$$= 132$$

$$\text{Failures, } B = (0.70)(440)$$

$$= 308$$

Employ Equation 13.14 to calculate χ^2 (see also Chapter 13):

$$\chi^2 = \frac{(128-132)^2}{132} + \frac{(312-308)^2}{308}$$

$$= 0.121 + 0.052$$

$$= 0.173$$

Because there are only two designs, there is only 1 degree of freedom. From Table 13.3, one notes that with 1 degree of freedom, there is only a 0.70 (70%) probability that χ^2 would be larger than 0.173. Therefore, the probability that there is no difference between the two pump designs is less than 70%.

APPLICATION 15.17

A new radiant heat transfer rod has been proposed for use in nuclear power plants. One of the first steps in determining the usefulness of this form of *hot* rod is to analytically estimate the temperature profile in the rod.

A recent environmental engineering graduate has proposed to estimate the temperature profile of the square pictured in Figure 15.3. She sets out to outline

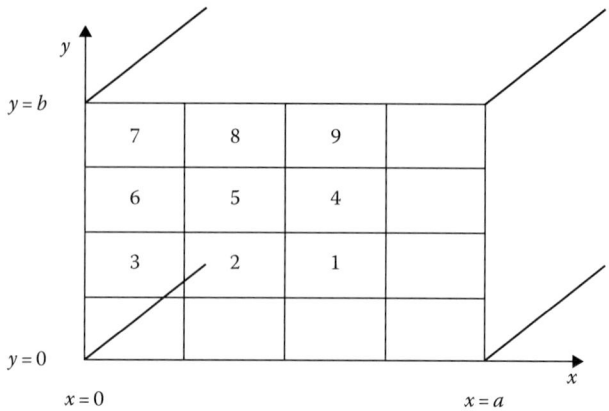

FIGURE 15.3   Monte Carlo grid (square surface).

a calculational procedure to determine the temperature profile. One of the options available is to employ the Monte Carlo method. Provide an outline of that procedure.[6]

Solution

One method of solution involves the use of the Monte Carlo approach employing the use of random numbers. Consider the square (it could also be a rectangle) pictured in Figure 15.3. If the describing equation for the variation of T within the grid structure is[6]

$$\frac{\partial^2 T}{\partial x^2} + \frac{\partial^2 T}{\partial y^2} = 0 \tag{15.4}$$

with specified boundary conditions (BC) for $T(x, y)$ of $T(0, y)$, $T(a, y)$, $T(x, 0)$, and $T(x, b)$, one may employ the following approach:

1. Proceed to calculate T at point 1, that is, T_1.
2. Generate a random number between 00 and 99.
3. If the number is between 00 and 24, move to the left. For 25–49, 50–74, and 75–99, move upward, to the right, and downward, respectively.
4. If the move in step 3 results in a new position that is *at* an outer surface (boundary), terminate the first calculation for point 1 and record the T value of the boundary at the new position. However, if the move results in a new position that is *not at* a boundary and is still at one of the nine internal grid points, repeat step 2 and step 3. This process is continued until an outer surface or boundary is reached.
5. Repeat step 2 to step 4 numerous times, e.g., 1000 times.
6. After completing step 5, sum all the T values obtained and divide this value by the number of times step 2 to step 4 have been repeated. The resulting value provides a reasonable estimate of T_1.
7. Return to step 1 and repeat the calculations for the remaining eight grid points.

REFERENCES

1. L. Theodore and R. Dupont, *Environmental Health and Hazard Risk Assessment: Principles and Calculations*, CRC Press/Taylor & Francis Group, Boca Raton, FL, 2013.
2. S. Shaefer and L. Theodore, *Probability and Statistics Applications in Environmental Science*, CRC Press/Taylor & Francis Group, Boca Raton, FL, 2007.
3. L. Theodore and F. Taylor, *Probability and Statistics*, Theodore Tutorials (originally published by USEPA, RTP, NC), East Williston, NY, 1996.
4. M.K. Theodore and L. Theodore, *Introduction to Environmental Management*, CRC Press/Taylor & Francis Group, Boca Raton, FL, 2010.
5. K. Skipke and L. Theodore, CRC Press/Taylor & Francis Group, Boca Raton, FL, 2014.
6. L. Theodore, *Heat Transfer Applications for the Practicing Engineer*, John Wiley & Sons, Hoboken, NJ, 2012.
7. L. Theodore, *Chemical Engineering: The Essential Reference*, McGraw-Hill, New York, NY, 2014.
8. J. Reynolds, J. Jeris, and L. Theodore, *Handbook of Chemical and Environmental Engineering Calculations*, John Wiley & Sons, Hoboken, NJ, 2004.
9. P. Abulencia and L. Theodore, *Fluid Flow for the Practicing Chemical Engineer*, John Wiley & Sons, Hoboken, NJ, 2010.
10. L. Theodore, *Air Pollution Control Equipment Calculations*, John Wiley & Sons, Hoboken, NJ, 2008.

16 Manufacturing and Electronics

There are 19 illustrative examples in this chapter. They key on environmental risk calculations that involve both discrete and continuous probability distributions. Discrete distributions that receive attention include binomial, multinomial, hypergeometric, and Poisson. Continuous distributions include: normal, log-normal, exponential, Weibull, chi-square (χ^2), F, and Monte Carlo. Two of the applications are based on activities at the international level.

A significant number of applications have been adapted from the literature.[1-3]

APPLICATION 16.1

Consider the case of a box of 100 transistors from an electronics firm from which a sample of two items is to be drawn *without* replacement. If the box contains five defective transistors, what is the probability that the sample contains exactly two defectives?

Solution

Let A denote the event that the first transistor drawn is defective, and B, the event that the second is defective. Then, the probability that the sample contains exactly two defectives is $P(AB)$. By application of the multiplication theorem provided in Equation 3.28 (from Chapter 3), one obtains

$$P(AB) = P(A)P(B|A)$$

$$P(AB) = \left(\frac{5}{100}\right)\left(\frac{4}{99}\right)$$

$$= 0.002 = 0.2\%$$

APPLICATION 16.2

Determine the reliability of the components A, D, and G of a manufacturing facility's electrical system illustrated in Figure 16.1. Use the reliabilities indicated under various components. The overall reliability has been determined to be 0.42.

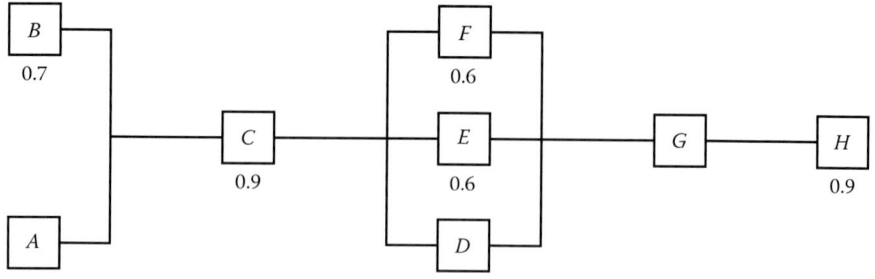

FIGURE 16.1   Diagram of an electrical system of Application 16.2.

Solution

The reliability of the parallel subsystem consisting of components A and B is obtained by applying Equation 11.8 (from Chapter 11), which yields

$$R_p = 1-(1-0.7)(1-A)$$
$$= 1-(-0.3-A+0.7A)$$
$$= 0.7+0.3A$$

The reliability of the parallel subsystem consisting of components D, E, and F is

$$R_p = 1-(1-D)(1-0.6)(1-0.6)$$
$$= 1-0.16(1-D)$$
$$= 0.84+0.16D$$

The reliability of the entire system is obtained by applying Equation 11.7 (from Chapter 11), which yields

$$R_s = 0.42 = (0.7+0.3A)(0.9)(0.84+0.16D)(G)(0.9)$$

The above single equation contains three unknowns. An infinite number of solutions are possible, including, for example,

$$A = 0; \quad D = 0; \quad G = 0.882$$

APPLICATION 16.3

Let the discrete random variable X represent the number of years before a furnace at a manufacturing facility is decommissioned because of environmental concerns. The pdf of X is given by

$$f(x) = \frac{k}{x}; \quad x = 1, 2, 3, 4$$

Find the constant k.

Solution

Assuming a valid pdf,

$$\sum_{1}^{4} f(x) = \sum_{1}^{4} \left(\frac{k}{x}\right) = 1.0; \quad x = 1,2,3,4$$

Substituting gives

$$k + \frac{k}{2} + \frac{k}{3} + \frac{k}{4} = 1.0$$

Solving for k,

$$k = \frac{12}{25} = 0.48$$

APPLICATION 16.4

The terms *fans* and *blowers* are often used interchangeably and are not differentiated in the following discussion: thus, any statements about fans apply equally to the blowers. Strictly speaking, however, fans are used for low-pressure (drop) operation, generally below 2.0 lb/in.^{2} (psi). Blowers are generally employed when generating pressure heads range from 2.0 to 14.7 psi. Higher-pressure operations require compressors.[4–6]

Fans are usually classified as centrifugal or axial-flow type. In *centrifugal fans*, the gas is introduced into the center of a revolving wheel (the eye) and discharged at right angles to the rotating blades. In *axial-flow fans*, the gas moves directly (forward) through the axis of rotation of the fan blades. Both types are used in industry, but it is the centrifugal fan that is employed at most facilities.[4–6]

A sample of 5 fans is drawn *with replacement* from a storage facility that contains 5% of defective fans. The term with replacement indicates that each fan tested is returned to the lot before the next is drawn. What is the probability that the number of defective fans in the sample is at most 2?

Solution

This random experiment consists of drawing a fan at random *with* replacement from a lot. The random experiment is performed 5 times because a sample of 5 fans is drawn with replacement from the lot. Therefore, $n = 5$. Also note that the performances are independent because each transistor is replaced before the next is drawn. Therefore, the composition of the lot is exactly the same before each drawing.

For this problem, once again associate *success* with drawing a defective fan. Associate *failure* with drawing a nondefective. Because 5% of the lot is defective, $p = 0.05$. Therefore, $q = 0.95$. Substitute these values (n, p, and q) in the binomial pdf.

$$f(x) = \frac{5!}{x!(5-x)!}(0.05)^x(0.95)^{5-x}; \quad x = 0, 1, \ldots, 5$$

For less than three defectives, x can assume values of 0, 1, and 2. One may now substitute the appropriate values of X to obtain the required probabilities.

$$P(\text{exactly 0 defectives}) = P(X = 0) = \frac{5!}{0!5!}(0.05)^0(0.95)^5$$

$$= 0.774 = 77.4\%$$

$$P(\text{exactly 1 defective}) = P(X = 1) = \frac{5!}{1!4!}(0.05)^1(0.95)^4$$

$$= 0.0204 = 2.14\%$$

$$P(\text{exactly 2 defectives}) = P(X = 2) = \frac{5!}{2!3!}(0.05)^2(0.95)^3$$

$$= 0.0214 = 2.14\%$$

Therefore,

$$P(X < 3) = P(X = 0) + P(X = 1) + P(X = 2)$$

$$= 0.774 + 0.204 + 0.0214$$

$$= 0.999 = 99.9\%$$

Is this a reasonable probability from a risk perspective?

APPLICATION 16.5

If 20% of the screws produced by a machine are environmentally defective, determine the probability that out of four screws chosen at random, 1 screw will be defective. Also calculate the probability that less than 2 screws will be defective. What is the probability that at least 2 screws will be defective?

Solution

The probability of a defective screw is $p = 0.2$ and of a nondefective screw is $q = 1 - p = 0.8$. Let the random variable X be the number of defective screws. Then,

$$P(X = 1) = \frac{4!}{1!3!}(0.2)^1(0.8)^3 = \binom{4}{1}(0.2)^1(0.8)^3$$

$$= 0.4096 = 41\%$$

Also,

$$P(X=0) = \binom{4}{0}(0.2)^0(0.8)^4$$

$$= 0.4096$$

$$P(X<2) = P(X=0) + P(X=1)$$

$$= 0.4096 + 0.4096$$

$$= 0.8192 = 81.92\%$$

Finally,

$$P(X \geq 2) = P(X=2) + P(X=3) + P(X=4)$$

$$= 1 - P(X<2)$$

$$= 1 - 0.8192$$

$$= 0.1808 = 18.08\%$$

Note that $P(X<2)$ and $P(X \geq 2) = 1,0$

APPLICATION 16.6

An engineer's ability to distinguish a normal manufacturing product from an environmentally unacceptable one (defective) is tested independently on 10 different occasions. What is the probability of eight correct identifications if the engineer is *only guessing*?

Solution

For *guessing* purposes, the probability of making a correct identification of the product may be reasonably assumed to be 0.5. Let X denote the number of correct identifications. If the engineer is only guessing, X has a binomial distribution with $n = 10$ and $p = 0.5$. The probability of exactly eight correct identifications is

$$P(X=8) = \frac{10!}{8!2!}(0.5)^8(0.5)^2$$

$$= 0.044 = 4.4\%$$

APPLICATION 16.7

A large bin contains an assortment of bolts to be employed in the manufacturing of small air conditioners for a quality control laboratory. Approximately, 50% of the bolts are square, 20% are hexagonal, and 30% are octagonal. If a sample of 5 bolts is drawn from the bin, what is the probability that 3 will be square, 2 will be hexagonal, and none will be octagonal?

Solution

This involves a multinomial distribution application with

$$p_1 = 0.5$$

$$p_2 = 0.2$$

$$p_3 = 0.3$$

Substitute into the multinomial equation; see also Equation 6.9 (from Chapter 6).

$$P(3,2,0) = \frac{5!}{3!2!0!}(0.5)^3 (0.2)^2 (0.3)^0$$

$$= 0.05 = 5\%$$

APPLICATION 16.8

Valves have two main functions in a pipeline: to control the amount of flow or to stop the flow completely. There are many different types of valves; the most commonly used are the gate valve and the globe valve. However, there are other valves.

Stop valves are used to shut off or, in some cases, partially shut off the flow of fluid. Stop valves are controlled by the movement of the valve stem. Stop valves can be divided into four general categories: globe, gate, butterfly, and ball valves. Plug valves and needle valves may also be considered stop valves.

Ball valves, as the name implies, are stop valves that use a ball to stop or state the flow of fluid. The ball performs the same function as the disk in the globe valve. When the valve handle is operated to open the valve, the ball rotates to a point where the hole through the ball is in line with valve body inlet and outlet. When the valve is shut, which requires only a 90° rotation of the handwheel for most valves, the ball is rotated so the hole is perpendicular to the flow openings of the valve body, and the flow is stopped.

As described in Chapter 7, the hypergeometric distribution is applicable in situations in which a random sample of n items is drawn without replacement from a set of N items. *Without replacement* means that an item is not returned to the set after it is drawn. Recall that the binomial distribution is frequently applicable in cases where the item is drawn with replacement.

Suppose that it is possible to classify each of the N items as a *success* or *failure*. Again, the words "success" and "failure" do not have the usual connotation. They are merely labels for two mutually exclusive categories into which N items have been classified.

Let a be the number of items in the category labeled "success." Then $N - a$ will be the number of items in the category labeled "failure." Let X denote the number of successes in a random sample of n items drawn without replacement from the set of N items. Then, the random variable X has a hypergeometric distribution whose probability distribution function (pdf) is specified as follows:

$$f(x) = \frac{\dfrac{a!}{x!(a-x)!} \dfrac{(N-a)!}{(n-x)!(N-a-n+x)!}}{\dfrac{N!}{n!(N-n)!}}; \quad x = 0,1,\ldots,\min{(a,n)} \quad (16.1)$$

The term $f(x)$ is the probability of x successes in a random sample of n items drawn without replacement from a set of N items, a of which are classified as successes and $N - a$ as failures. The term $\min(a, n)$ represents the smaller of the two numbers a and n, that is, $\min(a,n) = a$ if $a < n$ and $\min(a,n) = n$ if $n \le a$.

A quality control engineer inspects a random sample of 3 ball valves drawn *without replacement* from each incoming lot of 25 ball valves. A lot is accepted only if the sample contains no defectives. What is the probability of accepting a lot containing 10 defective ball valves?

Solution

Refer to Equation 16.1. Each lot consists of 25 valves. Therefore, $N = 25$. For this problem, $a = 10$. A sample of 3 ball valves is drawn without replacement from each lot; therefore, $n = 3$.

Substitute the values of N, n, and a in the pdf of the hypergeometric distribution, that is, Equation 16.1.

$$f(x) = \frac{\dfrac{10!}{x!(10-x)!} \dfrac{15!}{(3-x)!(12+x)!}}{\dfrac{25!}{(3!)(22)!}}; \quad x = 0,1,2,3$$

Substitute the appropriate values of X to obtain the required probability. The probability of accepting a lot containing 10 defective ball valves is

$$P(X=0) = \frac{\dfrac{10!}{(0!)(10!)} \dfrac{15!}{(3-0)!(12!)}}{\dfrac{25!}{(3!)(22)!}}$$

$$= \frac{(1)(455)}{2300}$$

$$= 0.20 = 20\%$$

APPLICATION 16.9

Consider a standby pumping redundancy system at a small coal-fired utility plant in a manufacturing facility in China with 1 operating unit and 3 on standby. (In effect, the system can survive three failures.) If the failure rate is estimated to be 1 unit per year, what is the 2-year reliability of the system?

Solution

This involves the application of the Poisson distribution. For this application,

$$\mu = (1)(2) = 2$$

The number of standby units is 3, that is, $n = 3$. Therefore,

$$R(x) = \sum_{0}^{3} \frac{e^{-2}2^x}{x!}$$

For this application,

$$
\begin{aligned}
R(X \le 3) &= R(X = 0) + R(X = 1) + R(X = 2) + R(X = 3) \\
&= e^{-2} + 2e^{-2} + e^{-2}(4/2) + e^{-2}(8/6) \\
&= 0.135 + 0.271 + 0.27 + 0.180 \\
&= 0.857 = 85.7\%
\end{aligned}
$$

APPLICATION 16.10

The average number of breakdowns of power supply at a manufacturing facility during 1000 days of operation is 6. What is the probability of no breakdowns during a 20-day work period?

Solution

The given average is the average number of breakdowns during 1000 months. The unit of time associated with the given average is 1000 days. The probability required is the probability of no breakdowns during a 20-day period. The unit of time connected with the required probability is therefore 20 days. In a 20-day period, there are 0.02, that is, 20/1000 time periods of 1000 days of duration. Since the given average is 6, multiplying 6 by 0.02 yields 0.12, the average number of occurrences during a 20-day time period. Substitute this values of μ into the Poisson pdf.

$$f(x) = \frac{e^{-0.12}(0.12)^x}{x!}; \quad x = 0, 1, 2, \dots$$

Substitute for x in the Poisson pdf the number of occurrences whose probability is required. The probability of no breakdowns in a 20-day period is

$$P(X = 0) = \frac{e^{-0.12}(0.12)^0}{0!} = e^{-0.12}$$

$$= 0.887 \approx 90\%$$

APPLICATION 16.11

The temperature of an estuary polluted from a manufacturing discharge during the winter months is normally distributed with a mean 46°F and a standard deviation 60°F. Calculate the probability that the temperature is between 45°F and 52°F.

Solution

Normalizing the temperature T gives

$$Z_1 = \frac{45 - 46}{3.0} = -0.333$$

$$Z_2 = \frac{52 - 46}{3.0} = 2.0$$

Refer to Table 9.1 (in Chapter 9). Therefore,

$$P(-0.333 < Z < 2.0) = P(0.0 < Z < 2.0) - P(-0.333 < Z < 0.0)$$

$$= 0.4722 - (-0.1293)$$

$$= 0.6015 = 60.15\%$$

APPLICATION 16.12

The lifetime, T, of a circuit board manufactured at an electronics facility for a nuclear power plant in Japan is normally distributed with a mean of 40 months. What is the largest lifetime variance the installed circuit boards can have if 95% of them need to last at least 12 months?

Solution

For this application,

$$P(T > 12) = 0.95$$

Normalizing gives

$$P\left(\left[\frac{T-\mu}{\sigma}\right] > \left[\frac{12-40}{\sigma}\right]\right) = 0.95$$

$$P\left(Z > -\frac{28}{\sigma}\right) = -1.645$$

The following equation must apply for this condition:

$$-\frac{28}{\sigma} = -1.64$$

Solving,

$$\sigma = 17\,\text{months}$$

APPLICATION 16.13

The data in Table 16.1 represent the time in microseconds between requests for a certain process service on a computer network employed at a uranium processing facility. Calculate the mean and standard deviation of the natural logs of observations presented in Table 16.1.

Solution

A frequency distribution is a table showing the number of observations in each of a succession of intervals called *classes*, as selected in Table 16.2.

Compute the mean, μ, and standard deviation, σ, of the natural logs of observations. Employ the results provided in Table 16.3. The mean, μ, is therefore

$$\mu = \sum_{i=1}^{50} \frac{\ln X_i}{50} = \frac{446.51}{50} = 8.93$$

TABLE 16.1

Computer Network Data

114,462	10,280	2,654	6,761	8,111
5,437	14,691	4,605	9,405	15,184
4,866	4,789	11,944	6,919	5,547
1,439	1,333	18,270	35,632	17,783
13,017	32,145	7,310	1,812	15,078
4,138	7,361	9,405	4,277	2,592
1,594	39,577	3,820	6,925	6,974
1,422	6,063	5,432	6,003	27,778
36,938	15,615	2,904	8,840	3,711
10,829	5,575	6,634	3,674	5,825

TABLE 16.2
Computer Network Frequency Distribution

Interrequest Time (µs)	Frequency
0–2,499	5
2,500–4,999	11
5,000–9,999	18
10,000–19,999	10
20,000–39,999	5
40,000–79,999	0
80,000–159,999	1
	Total = 50

TABLE 16.3
Natural Log Frequency

X	ln X	X	ln X	X	ln X
5,437	8.60	6,063	8.71	1,812	7.5
4,866	8.50	15,615	9.66	4,277	8.36
1,439	8.49	5,575	8.63	6,925	8.84
13,017	9.47	2,654	7.88	6,003	8.70
4,138	8.33	4,605	8.43	8,840	9.09
1,594	7.37	11,944	9.39	3,674	8.21
114,462	11.65	18,270	9.81	8,111	9.00
1,422	7.26	7,310	8.90	15,184	9.63
36,938	10.52	7,405	9.15	5,547	8.62
10,829	9.29	3,820	8.25	17,783	9.79
10,280	9.24	5,432	8.60	15,078	9.62
14,691	9.59	2,904	7.97	2,592	7.86
4,789	8.47	6,634	8.80	6,974	8.85
1,333	7.20	6,761	8.82	27,778	10.23
32,145	10.38	9,405	9.15	3,711	8.22
7,361	8.90	6,919	8.84	5,825	8.67
39,577	10.59	35,632	10.48		

Estimate of $\alpha = 8.93$.
Estimate of $\beta = 0.91$.

The standard deviation is

$$\sigma = \sqrt{\sum_{i=1}^{50} \frac{[\ln X_i - 8.93]^2}{49}} = 0.91$$

Estimate α and β by the mean and the standard deviation obtained in Table 16.3.

$$\alpha = 8.93$$
$$\beta = 0.91$$

Was this example presented earlier? The reader may choose to return to Chapter 11 and examine Illustrative Example 10.2.

APPLICATION 16.14

A pumping system at a water treatment facility at an electronics plant consists of 4 components. Three are connected in parallel, which in turn are connected downstream (in series) with the other component. The arrangement is schematically shown in Figure 16.2. If the pumps have the same exponential failure rate, λ, of 0.30 (year)$^{-1}$, estimate the probability that the system will not survive for more than two years.

Solution

Based on the information provided, the pumping system fails when the three parallel components fail or when the downstream component fails. This is a combination of a parallel and series systems. From Equation 11.3 (in Chapter 11), the reliability is

$$R(t) = 1 - F(t)$$

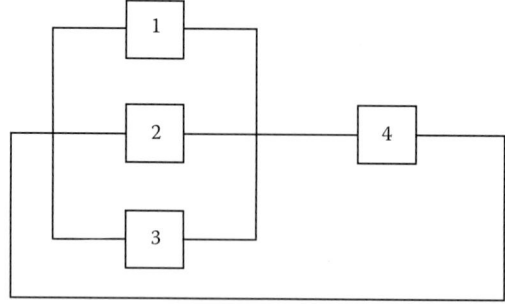

FIGURE 16.2   Pumping system in Application 16.14.

where $F(t)$ is the probability of failure between 0 and t. From Equation 11.8 (in Chapter 11), the reliability of the parallel system is

$$R_p = 1 - (1 - R_1)(1 - R_2)(1 - R_3); R_i = R$$
$$= 1 - (1 - R)^3$$

From Equation 11.7 (in Chapter 11), for a series system

$$R_s = R_p R_4 = R_p R$$

where R_s also represents the overall system reliability.
   Applying the exponential model in Equation 11.3 (in Chapter 11) gives

$$R(t) = 1 - F(t)$$
$$= 1 - (1 - e^{-\lambda t})$$
$$= e^{-\lambda t}$$

$$R_1 = R_2 = R_3 = R_4 = R = e^{-(0.30)(2)} = e^{-0.6}$$
$$R = 0.549$$

Thus,

$$R_p = 1 - (1 - 0.549)^3$$
$$= 1 - 0.092$$
$$= 0.918 = 91.8\%$$

Since

$$R_4 = 0.549$$

Then,

$$R_s = (0.918)(0.549)$$
$$= 0.504 = 50.4\%$$

APPLICATION 16.15

The life (time to failure) of a machine component has a Weibull distribution. Outline how to determine the probability that the component lasts a given period of time if the failure rate is $t^{-1/2}$ over its entire time domain.

Solution

As presented in Illustrative Example 12.1 (in Chapter 12), $\alpha = 2$ and $\beta = 1/2$. For these values of α and β given earlier, the Weibull pdf is

$$f(t) = t^{-1/2} e^{-2t^{1/2}}; \quad t > 0$$

Integration of this pdf will yield the required probability. This integration is detailed in several illustrative examples in earlier and later chapters.

APPLICATION 16.16

Past experience in a manufacturing facility indicates that electrical wires in an electrostatic precipitate purchased from a company have a mean break strength of 40.0 lb and a standard deviation of 1.5 lb.

1. If 16 rods are selected, between what two values could one reasonably expect their mean to be?
2. How many rods need to be selected so that one would be certain with a probability of 95% (0.95) that the resulting mean would not be in error by more than 0.2 lb?

Solution

1. Note: Since this involves a two-tailed problem, and the problem merely states *reasonably expect*, one may arbitrarily define the 95% confidence limits as those two values between which one can expect the mean to be, i.e.,

$$P(-1.96 < Z < 1.96) = 0.95$$

First note that

$$\sigma = \frac{1.5}{\sqrt{16}} = 0.375$$

Therefore,

$$-1.96 < \frac{\bar{X} - \mu}{\sigma} < 1.96$$

$$-1.96 < \frac{\bar{X} - 40.0}{0.375} < 1.96$$

$$39.265 < \bar{X} < 40.735$$

2. For this calculation,

$$P(39.8 < \bar{X} < 40.2) = 0.95$$

Once again

$$P\left(\frac{39.8 - 40.0}{1.5/\sqrt{n}} < Z < \frac{40.2 - 40.0}{1.5/\sqrt{n}}\right) = 0.95$$

$$1.96 = \frac{0.2}{\left(1.5/\sqrt{n}\right)}$$

$$n = 216$$

APPLICATION 16.17

A contingency table is a classification of observations according to two or more criteria of classification. An $r \times c$ contingency table features r rows and c columns. Contingency tables are used to test the independence of the criteria of classification. The observed frequency of each cell, that is, the intersection of a row and column is independent; multiplying the row total by the column total and then dividing the result by the grand total lead to the theoretical frequency of each cell. The chi-square test statistic is then computed. Under the assumption of independence, the chi-square test is approximately distributed as a random variable having a chi-square distribution with $(r-1)(c-1)$ degrees of freedom. Large values of the chi-square test statistic lead to the rejection of the assumption of independence.

A sample of 300 items from one day's manufacturing production of a factory is classified as to work shift and quality as follows:

Quality	Day	Evening	Midnight	Total
Poor	12	12	10	34
Satisfactory	90	68	72	230
Excellent	21	10	5	36
Total	123	90	87	300

Test the assumption that the quality of items produced is independent of the work shift in which they are produced. Use a 5% level of significant.[1,2]

Solution

The product of the row total and the column total divided by the grand total is computed for each cell:

$$e_{11} = \frac{(34)(123)}{300} = 13.9$$

$$e_{12} = \frac{(34)(90)}{300} = 10.2$$

$$e_{13} = \frac{(34)(87)}{300} = 9.9$$

$$e_{21} = \frac{(230)(123)}{300} = 94.3$$

$$e_{22} = \frac{(230)(90)}{300} = 69.0$$

$$e_{23} = \frac{(230)(87)}{300} = 66.7$$

$$e_{31} = \frac{(36)(123)}{300} = 14.8$$

$$e_{32} = \frac{(36)(90)}{300} = 10.8$$

$$e_{33} = \frac{(36)(87)}{300} = 10.4$$

Note that e_{ij} represents the theoretical frequency for the cell at the intersection of the ith row and the jth column.
The test statistic is given by

$$\sum_{i=1}^{r}\sum_{j=1}^{c} \frac{(f_{ij} - e_{ij})^2}{e_{ij}}$$

where
f_{ij} is the observed frequency of the ith row and jth column, that is, the cell at the intersection of the ith row and the jth column
e_{ij} is the corresponding theoretical frequency
r is the number of rows
c is the number of columns

Therefore, the chi-square test statistic is

$$\chi^2 = \frac{(12-13.9)^2}{13.9} + \frac{(12-10.2)^2}{10.2} + \frac{(10-9.9)^2}{9.9} + \frac{(90-94.3)^2}{94.3} + \frac{(68-69)^2}{69}$$
$$+ \frac{(72-66.7)^2}{66.7} + \frac{(21-14.8)^2}{14.8} + \frac{(10-10.8)^2}{10.8} + \frac{(5-10.4)^2}{10.4}$$
$$= 6.67$$

The probability (P-value) that a chi-square variable with $(r - 1)(c - 1)$ degrees of freedom exceeds the value of the chi-square test statistic from Table 13.5 (from Chapter 13) is

$$P\left(\chi^2_{(r-1)(c-1)} > 6.67\right) > 0.10; \quad r = 3, c = 3; \quad (r-1)(c-1) = (2)(2) = 4$$
$$P\left(\chi^2_4 > 6.67\right) > 0.10$$

Because the P-value exceeds 0.10, one would fail to reject the assumption that the quality of an item is independent of the work shift during which it was produced. Note that rejection of independence implies that the row and column criteria of classification are interrelated.

APPLICATION 16.18

The following data are provided on two sets, $n_1 = 10$, $n_2 = 15$, of electrical wire defects from populations with equal variances, that is, $\sigma_1 = \sigma_2$. Calculate

$$P\left(s_1^2 / s_2^2 < 2.65\right)$$

Solution

For this one-sided, right-handed test,

$$P\left(s_1^2 / s_2^2 < 2.65\right) = P\left(F < 2.65\right)$$

with $v_1 = 9$ and $v_2 = 14$. From Table 13.5;

$$P\left(F < 2.65\right) = 1 - 0.05 = 0.95$$

APPLICATION 16.19

A series system in a manufacturing facility in China consists of two electrical components, A and B. Component A has a time to failure, T_A, assumed to be normally distributed with a mean (μ) 100 months and a standard deviation (σ) 20 months. Component B has a time to failure, T_B, assumed to be normally distributed with a mean 90 months and a standard deviation 10 months. The system fails whenever either component A or component B fails. Therefore, T_S, the time to failure of the system is the minimum of the times to failure of components A and B.

Estimate the average value of T_S on the basis of the simulated values of 10 simulated values of T_A and 10 simulated values of T_B.

Solution

First, generate 20 random numbers in the range of 0–1[1,2]:

$$0.10, \ 0.54, \ 0.42, \ 0.02, \ 0.81, \ 0.07, \ 0.06, \ 0.27, \ 0.57, \ 0.80,$$
$$0.92, \ 0.86, \ 0.45, \ 0.38, \ 0.88, \ 0.21, \ 0.26, \ 0.51, \ 0.73, \ 0.71$$

Use the standard normal distribution table in Chapter 9 and obtain the simulated value of Z corresponding to each of the random numbers. The first random number is 0.10. The corresponding simulated value of Z is –1.28 because the area under a standard normal curve to the left of –1.28 is 0.10. The remaining simulated values of Z are obtained in a similar fashion. The 20 simulated values of Z are provided in Table 16.4.

TABLE 16.4
Simulated Values of Z

Random Number	Simulated Values of Z
0.10	–1.28
0.54	0.10
0.42	–0.20
0.02	–2.05
0.81	0.88
0.07	–1.48
0.06	–1.56
0.27	–0.61
0.57	0.18
0.80	0.84
0.92	1.41
0.86	1.08
0.45	–0.13
0.38	–0.31
0.88	1.17
0.21	–0.81
0.26	–0.64
0.51	0.03
0.73	0.61
0.71	0.56

Using the first 10 simulated values of Z, obtain 10 simulated values of T_A by multiplying each simulated value of Z by 20 (the standard deviation, σ) and adding 100 (the mean, μ), that is,

$$T_A = \sigma Z + 100$$

Since

$$Z = \frac{T_A - 100}{\sigma}$$

Note that the lifetime or time to failure of each component, T, is calculated using this equation. Thus, multiplying each of the first 10 simulated values of Z by 20 and adding 100 yields the following simulated values of T_A:

$$74, \ 102, \ 96, \ 59, \ 118, \ 90, \ 69, \ 88, \ 104, \ 117$$

Multiplying each of the second 10 simulated values of Z by 10 and adding 90 yields the following simulated values of T_B:

$$104, \ 101, \ 89, \ 87, \ 102, \ 82, \ 84, \ 90, \ 96, \ 96$$

Simulated values of T_S corresponding to each pair of simulated values of T_A and T_B are obtained by recording the minimum of each pair. The values are shown in Table 16.5.

The average of the 10 simulated values of T_S is 84 months, the estimated time to failure of the system.

TABLE 16.5
Minimum Simulated Values

Simulated Time to Failure (months)

Component A (T_A)	Component B (T_B)	System (T_S)
74	104	74
102	101	101
96	89	89
59	87	59
118	102	102
70	82	70
69	84	69
88	90	88
104	96	96
117	96	96

REFERENCES

1. L. Theodore and F. Taylor, *Probability and Statistics*, Theodore Tutorials (originally published by USEPA, RTP, NC), East Williston, NY, 1996.
2. S. Shaefer and L. Theodore, *Probability and Statistics Applications in Environmental Science*, CRC Press/Taylor & Francis Group, Boca Raton, FL, 2007.
3. L. Theodore and R. Dupont, *Environmental Health and Hazard Risk Assessment: Principles and Calculations*, CRC Press/Taylor & Francis Group, Boca Raton, FL, 2013.
4. P. Abulencia and L. Theodore, *Fluid Flow for the Practicing Chemical Engineer*, John Wiley & Sons, Hoboken, NJ, 2010.
5. J. Reynolds, J. Jeris, and L. Theodore, *Handbook of Chemical and Environmental Engineering Calculations*, John Wiley & Sons, Hoboken, NJ, 2004.
6. L. Theodore, *Chemical Engineering: The Essential Reference*, McGraw-Hill, New York, NY, 2014.

17 Pharmaceuticals

There are 18 applications in this chapter. They key on environmental risk calculations that involve both discrete and continuous probability distributions. Discrete distributions that receive attention include: binomial, multinomial, hypergeometric, and Poisson. Continuous distributions include: normal, log-normal, exponential, Weibull, and *F*. Two of the problems are based on activities at the international level.

The bulk of the applications in this chapter has been adapted from the literature.[1–3]

APPLICATION 17.1

Consider the case of a box of 500 pills from which a sample of 10 pills is to be drawn *without replacement*. If the box contains 10 contaminated pills, what is the probability that the sample contains exactly 2 contaminated pills?

Solution

Let *A* denote the event that the first pill drawn is contaminated and *B*, the event that the second is contaminated. Then, the probability that the sample contains exactly 2 contaminated pills is *P(AB)*. By application of the multiplication theorem, one obtains (see also Equation 3.28 [in Chapter 3])

$$P(AB) = P(A)P(B|A)$$

$$P(AB) = \left(\frac{10}{500}\right)\left(\frac{9}{499}\right)(0.2)(0.018)$$

$$= 0.00361 = 0.36\%$$

APPLICATION 17.2

Suppose that an explosion at a pharmaceutical plant could have occurred as a result of one of three mutually exclusive human causes: stupidity, carelessness, or laziness. It is estimated that such an explosion could occur with a probability 0.20 as a result of stupidity, 0.40 as a result of carelessness, and 0.75 as a result of laziness. It is also estimated that the prior probabilities of the three possible causes of the explosion are, respectively, 0.50, 0.35, and 0.15. Using Bayes' theorem, determine the most likely cause of the explosion.

Solution

Let A_1, A_2, A_3 denote, respectively, the events in which stupidity, carelessness, and laziness are the problem. Let B denote the event of the explosion. Then,

$$P(A_1) = 0.50; \quad P(B|A_1) = 0.20$$

$$P(A_2) = 0.35; \quad P(B|A_2) = 0.40$$

$$P(A_3) = 0.15; \quad P(B|A_3) = 0.75$$

Applying Bayes' theorem,

$$P(A_1 \mid B) = \frac{P(A_1)P(B|A_1)}{P(A_1)P(B|A_1) + P(A_2)P(B|A_2) + P(A_3)P(B|A_3)}$$

and substituting,

$$P(A_1|B) = \frac{(0.50)(0.20)}{(0.50)(0.20) + (0.35)(0.40) + (0.15)(0.75)} = 0.28$$

Similarly,

$$P(A_2|B) = 0.40; \quad P(A_3 \mid B) = 0.32$$

Therefore, carelessness is the most likely cause of the explosion.

APPLICATION 17.3

Let X denote the annual number of new drugs developed by a pharmaceutical company. The probability distribution function (pdf) of X is specified as

$$f(x) = 0.25; \quad x = 0$$

$$f(x) = 0.35; \quad x = 1$$

$$f(x) = 0.24; \quad x = 2$$

$$f(x) = 0.11; \quad x = 3$$

$$f(x) = 0.04; \quad x = 4$$

$$f(x) = 0.01; \quad x = 5$$

1. What is the probability of developing 4 or more new drugs in a year, given that 2 drugs have already been developed?
2. What is the probability of at least 1 more drug is developed in a year, given that 1 has already been developed?

Solution

1. This involves a conditional probability calculation. The describing equation and solution is

$$P(X \geq 4 \backslash X \geq 2) = P(X \geq 4 \text{ and } X \geq 2)/P(X \geq 2)$$

$$= P(X \geq 4)/P(X \geq 2)$$

$$= 0.05/0.40$$

$$= 0.125 = 12.5\%$$

2. This also involves a conditional probability calculation. For this case,

$$P(X \geq 2 \backslash X \geq 1) = P(X \geq 2 \text{ and } X \geq 1)/P(X \geq 1)$$

$$= P(X \geq 2)/P(X \geq 1)$$

$$= 0.40/(0.40 + 0.35)$$

$$= 0.40/0.75$$

$$= 0.53 = 53\%$$

APPLICATION 17.4

Determine the reliability of the components A, D, and G at the pharmaceutical plant illustrated in Figure 17.1. Use the reliabilities indicated under various components. The overall reliability has been determined to be 0.42. Determine the reliability of component D if $A = 0.7$.

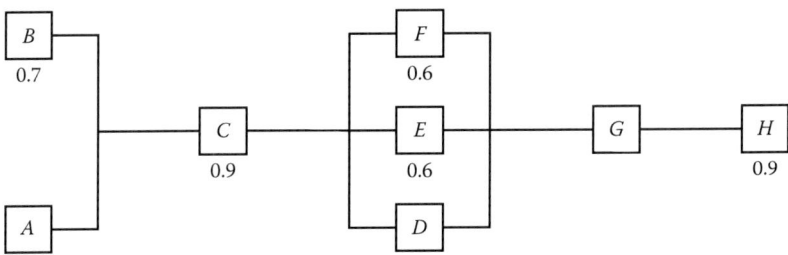

FIGURE 17.1 Diagram of a pharmaceutical plant.

Solution

With $A = 0.7$, the equation for R_s (see Equation 11.8) becomes

$$R_s = 0.42 = (0.91)(0.9)(0.84 + 0.16D)(G)(0.9)$$

There is one equation and two unknowns. An infinite number of solutions are possible, including, for example,

$$D = 0; \quad G = 0.678$$

APPLICATION 17.5

Refer to Application 17.4. Determine the reliability of component G if $A = 0.7$ and $D = 0.09$.

Solution

Here,

$$R_s = 0.42 = (0.91)(0.9)(0.84 + 0.16(0.09))(G)(0.9)$$

Solving for G,

$$G = 0.667$$

APPLICATION 17.6

A sample of 5 drugs is drawn *with replacement* from a bin in which 5% are contaminated. Once again, *with replacement* means that each drug drawn is returned to the bin before the next is drawn. What is the probability that the number of contaminated drugs in the sample is exactly 4?

Solution

This random experiment consists of drawing a drug at random with replacement from a bin. This is a random experiment performed 5 times, that is, a sample of 5 drugs is drawn with replacement from the bin. Therefore, $n = 5$. Also note that the operations are independent because each drug is *replaced* before the next is drawn. Therefore, the composition of the bin is exactly the same before each drawing.

For this problem, associate *success* with drawing a defective drug. Associate *failure* with drawing a contaminated drug. Because 5% of the lot is defective, $p = 0.05$. Therefore, $q = 0.95$. Substitute these values (n, p, and q) in the binomial pdf:

$$f(x) = \frac{5!}{x!(5-x)!}(0.05)^x (0.95)^{5-x}; \quad x = 0, 1, \ldots, 5$$

Substitute the appropriate value of X to obtain the required probabilities:

$$P(\text{exactly 4 defectives}) = P(X = 4) = \frac{5!}{4!1!}(0.05)^4(0.95)^1$$

$$= 0.000030 = 0.003\%$$

As expected, the probability is extremely low.

APPLICATION 17.7

Refer to Application 17.6. A sample of 5 drugs is again drawn *with replacement* from a bin in which 5% of the drugs are contaminated. What is the probability that the number of contaminated drugs in the sample is

1. At most 2
2. At least 2

Solution

Employ the same approach as the previous application.

1. $P(\text{at most 2 contaminated drugs}) = P(X \leq 2)$. Therefore, substitute $X = 0$, 1 and into

$$P(X \leq 2) = \frac{5!}{x!(5-x)!}(0.05)^x(0.95)^{5-x}$$

$$= 0.9988 = 99.88\%$$

2. $P(\text{at least 2 contaminated drugs}) = P(X \geq 2)$. Therefore, substitute $X = 0$ and 1 into

$$P(X \geq 2) = 1 - P(X \leq 1)$$

$$= 1 - \frac{5!}{x!(5-x)!}(0.05)^x(0.95)^{5-x}$$

$$= 0.226 = 22.6\%$$

APPLICATION 17.8

The probabilities of hospitals A, B, and C obtaining a particular type of serum are 0.5, 0.3, and 0.2, respectively. Four such serums are to be provided. What is the probability that a single hospital receives all 4 serums?

Solution

This involves the use of the multinomial distribution. The desired probability is (see Equation 6.9 [in Chapter 6])

$$P(1 \text{ hospital alone getting serum}) = f(4,0,0) + f(0,4,0) + f(0,0,4)$$

$$= \frac{4!}{4!0!0!}(0.5)^4(0.3)^0(0.2)^0 + \frac{4!}{0!4!0!}(0.5)^0(0.3)^4(0.2)^0 + \frac{4!}{0!0!4!}(0.5)^0(0.3)^0(0.2)^4$$

$$= 0.0625 + 0.0081 + 0.0016$$

$$= 0.0722 = 7.22\%$$

Note that for this calculation, each hospital is assumed to receive all four shipments.

APPLICATION 17.9

A quality control engineer inspects a random sample of 5 similar drugs drawn *without replacement* from each lot of 50 of these drugs. A lot is accepted only if the sample contains *no*, that is, 0, contaminated drugs. What is the probability of accepting a lot containing 20 contaminated drugs (pills)?

Solution

This random variable X has a hypergeometric distribution whose pdf is specified as follows:

$$f(x) = \frac{\dfrac{a!}{x!(a-x)!}\dfrac{(N-a)!}{(n-x)!(N-a-n+x)!}}{\dfrac{N!}{n!(N-n)!}}; \quad x = 0,1,\ldots,\min(a,n) \quad (17.1)$$

The term $f(x)$ is the probability of x successes in a random sample of n items drawn without replacement from a set of N items, a of which are classified as successes and $N - a$ as failures.[1,2]

Each lot consists of 50 pills. Therefore, $N = 50$. For this problem, $a = 20$. Since a sample of 5 pills is drawn without replacement from each lot, $n = 5$.

Substitute the values of N, n, and a in the pdf of the hypergeometric distribution, that is, Equation 17.1.

$$f(x) = \frac{\dfrac{20!}{x!(20-x)!}\dfrac{30!}{(5-x)!(35+x)!}}{\dfrac{50!}{(5!)(45!)}}; \quad x = 0,1,2,3,4,5$$

Substitute the appropriate values of X to obtain the required probability. The probability of accepting a lot containing 10 defective ball valves is

$$P(X=0) = \frac{\dfrac{20!}{(0!)(20!)} \dfrac{30!}{(5-0)!(25!)}}{\dfrac{25!}{(3!)(22!)}}$$

$$= \frac{(1)(142,506)}{2,118,760}$$

$$= 0.067 = 6.7\%$$

APPLICATION 17.10

Suppose 30 minor accidents occur at a pharmaceutical's 50 international plants. Find the probability P that exactly 2 of these accidents occur at a given plant.

Solution

The number of accidents at one plant may be viewed as the number of successes. Here $n=30$ since there are 3 accidents, and $P=1/50$, the probability that an accident occurs appears at a given plant. Since P is small, the Poisson approximation to the binomial distribution applies with

$$\lambda = nP = 30(1/50) = 0.6$$

Substituting into the Poisson equation,

$$P = P(X=2) = \frac{(0.6)^2 e^{-0.6}}{2!} = \frac{(0.36)(0.549)}{2}$$

$$= 0.0988 \cong 10\%$$

APPLICATION 17.11

Consider the problem of determining whether or not the occurrence of 8 cases of leukemia among 7076 children in Nassau County is so unexpectedly large as to warrant further study by public health officials. An examination of public health records in Nassau County (home of the author) reveals that in the 2008–2013 period there were 286 diagnosed cases of leukemia out of 1,152,685 children.[4]

Solution

Thus, the probability that a child will be stricken with leukemia is $P=286/1,152,695=0.000248$. The probability of the occurrence of 8 or more cases of leukemia in a population of 7076 children may be calculated employing the

Poisson distribution. In a population of size n, the number of occurrences of leukemia is a binomially distributed random variable X with parameters $n=7076$, $P=0.000248$. Note that

$$P(X \geq 8) = 1 - \sum_{0 \leq x \leq 7} \binom{7076}{x} (0.000248)^x (0.999752)^{7076-x}$$

However, the Poisson distribution may be employed. For this case, $nP = 7076 \times 0.000248 = 1.75 = \lambda$. Therefore,

$$P(X \leq 7) \approx \sum_{0 \leq x \leq 7} e^{-1.75} \frac{1.75^x}{x!} = 0.999518$$

Consequently,

$$P(X \geq 8) \approx 1 - 0.999518$$

$$= 0.000482 = 0.04822\%$$

APPLICATION 17.12

The regulatory specification on a toxic in a solid waste discharge from a pharmaceutical plant calls for a level of 2.0 ppm or less. Earlier observations of the toxic concentration, C, of the waste indicate a normal distribution with a mean of 0.40 ppm and a standard deviation of 0.08 ppm. Estimate the probability that waste will exceed the regulatory limit.

Solution

This problem requires the calculation of $P(C > 2.0)$. Normalizing the variable C,

$$P\{[(C-0.4)/0.08] > [(2.0-0.4)/0.08]\}$$

$$P(Z > 2.0)$$

From the standard normal table,

$$P(Z > 2.0) = 0.0228$$

$$= 2.28\%$$

For this situation, the area to the right of 2.0 is 2.28% of the total area. This represents the probability that waste will exceed the regulatory limit of 2.0 ppm.

APPLICATION 17.13

Discuss some of the *medical* applications associated with the log-normal probability distribution.[4]

Solution

The log-normal distribution has been employed to characterize occupational expo-sures over time. It typically models a distribution of data where there are many mea-surements with lower values and a few measurements with much higher values. For example, this distribution can adequately model a workplace condition where there is a relatively stable background concentration that is occasionally punctuated by higher concentrations due to cyclic production steps and/or short-term tasks. This model is not perfect, but it is useful for evaluating exposures if one keeps in mind the underlying reasons for the exposure profiles.

The statistical description of an exposure profile is also often needed in epide-miology studies, particularly for dose–response analysis. Depending on whether a chemical has primarily acute or chronic toxicological effects, the epidemiologist might be most interested in the peak exposures of an exposure group or possibly the average exposures of an exposure group.

APPLICATION 17.14

Normalized biological oxygen demand (BOD) levels in an estuary downstream from a pharmaceutical plant in India during the past 10 years are summarized in Table 17.1. If the BOD levels are assumed to follow a log-normal distribution, predict the level that would be exceeded once in 40 years.

Solution

For this case, refer to Table 17.2. Based on the data presented in Table 17.2,

$$\bar{X} = \frac{\sum X}{n} = \frac{38.86}{10} = 3.886$$

$$s^2 = \frac{\sum X^2 - \left(\sum X\right)^{\frac{2}{n}}}{n-1}$$

$$= \frac{156.78 - \left(38.86\right)^2 / 10}{10-1}$$

$$= 0.64$$

TABLE 17.1

Estuary BOD Data

Year	1	2	3	4	5	6	7	8	9	10
BOD level	23	8	17	210	62	142	43	29	71	31

Note: BOD, biological oxygen demand.[5]

TABLE 17.2
BOD Calculations

Year (Y)	BOD Level	$X = \ln$ BOD	X^2
1	23	3.13	9.83
2	38	3.64	13.25
3	17	2.83	8.01
4	210	5.35	28.62
5	62	4.13	17.06
6	142	4.96	24.60
7	43	3.76	14.14
8	29	3.37	11.36
9	71	4.26	18.15
10	31	3.43	11.76
Total	—	38.86	156.78

---

and

$$s = 0.80$$

For this test, with $Z = 1.96$ for 97.5% (once in 40 years) value,

$$Z = \frac{X - \bar{X}}{s}$$

$$1.96 = \frac{X - 3.886}{0.80}$$

Solving for X yields

$$X = 5.454$$

Thus, for this log-normal distribution,

$$X = \ln (BOD)$$

$$X = 5.454 = \ln (BOD)$$

$$BOD = 334$$

APPLICATION 17.15

The probability that a thermometer in a batch reactor at a pharmaceutical plant will not survive for more than 1 year is 0.25; how often should the thermometer be replaced? Assume that the time to failure is exponentially distributed and that the replacement time should be based on the thermometer's expected life.

Solution

This requires the calculation of μ in the exponential model with units of (year)$^{-1}$. For this case,

$$F(t) = 1 - e^{-\lambda t}$$

Based on the information provided,

$$P(T \leq 1) = 0.25$$

or

$$0.25 = 1 - e^{-(\lambda)(1)}$$

Solving for λ gives

$$\lambda = -\frac{1}{1} \ln(1 - 0.25)$$

$$= 0.288 \ (year)^{-1}$$

Because the expected time (or life), $E(T)$, is

$$E(T) = \frac{1}{\lambda}$$

$$= \frac{1}{0.288}$$

$$= 3.48 \ years$$

the thermometer should therefore be replaced in approximately 3.5 years.

APPLICATION 17.16

The life span of a carcinogenic drug is a random variable having a Weibull distribution with $\alpha = 0.025$ and $\beta = 0.50$ (with time in days). A medical intern has reported that the life span is approximately 400–500 days. Comment on her conclusion.

Solution

Let T denote the life in days of the drug. The pdf of T is obtained by applying Equation 12.3, which yields

$$f(t) = \alpha \beta t^{\beta-1} \exp(-\alpha t^\beta); \quad t > 0; \quad \alpha > 0, \ \beta > 0$$

Substituting $\alpha=0.025$ and $\beta=0.50$ yields

$$f(t) = (0.025)(0.50)t^{0.50-1}\exp\left(-0.025t^{0.50}\right); t > 0$$

$$= (0.0125)t^{-0.50}\exp\left(-0.025t^{0.50}\right); t > 0$$

The average value for this continuous variable T is given by the integration of Equation 4.22 (in Chapter 4).

$$E(T) = \int_{-\infty}^{\infty} t\, f(t)\, dt$$

$$= \int_{-\infty}^{\infty} (0.0125)t^{-0.50}\exp\left(-0.025t^{0.50}\right)dt$$

Unlike many of the simple integrations presented earlier, this more complicated integral was solved employing an incomplete Gamma function. The effort produced the following result[5]:

$$E(T) = 3200 \text{ d}$$

The intern has apparently erred in her analysis.

APPLICATION 17.17

Table 17.3 shows the observed frequency distribution and theoretical frequency distribution of 50 interrequest times for computer service at a pharmaceutical plant in Canada. The theoretical frequencies were obtained under the assumption that inter request time for computer service has a log-normal distribution. Refer to Chapter 10 for additional details.

On the basis of the given data, test the assumption that interrequest time for computer service has a log-normal distribution.

TABLE 17.3
Frequency of Interrequest Time

Interrequest Time (μs)	Frequency	Theoretical Frequency
0–2,499	5	5.6
2,500–4,999	11	10.8
5,000–9,999	18	14.8
10,000–19,999	10	11.8
20,000–39,999	5	5.5
40,000–79,999	0	1.4
80,000–159,999	1	0.3

Solution

The number of classes in the frequency distribution, k, is 7. There are two parameters that can be estimated from the data, μ and σ. Therefore, $m=2$.

The degree of freedom, v, is therefore

$$v = k - m - 1$$

$$= 7 - 2 - 1 = 4$$

The chi-square test statistic is now evaluated employing Equation 13.14 (in Chapter 13).

$$\sum_{i=1}^{k} \frac{(f_i - e_i)^2}{e_i}$$

$$= \frac{(5-5.6)^2}{5.6} + \frac{(11-10.8)^2}{10.8} + \frac{(10-11.8)^2}{11.8} + \frac{(5-5.5)^2}{5.5} + \frac{(0-1.4)^2}{1.4} + \frac{(1-0.3)^2}{0.3}$$

$$= 4.83$$

The probability that a chi-square variable with v ($v=4$) degrees of freedom exceeds the observed value of the test statistic is given by (see Table 13.5 [in Chapter 13])

$$P\left(\chi^2 > 4.83\right) > 0.30$$

If the probability computed is less than 0.05, one would reject the assumption being tested; otherwise, one would fail to reject it. In this case, one would fail to reject the assumption that the interrequest time for computer service has a log-normal distribution. Note that rejecting the assumption being tested indicates that the data do not provide sufficient evidence in favor of the assumed distribution of the random variable.

APPLICATION 17.18

Five weight measurements from an analytical balance in a pharmaceutical laboratory produce results for the sample variance of $s_1^2 = 0.50$ mg. Ten weight measurements with another analytical balance produced a sample variance $s_2^2 = 0.10$ mg. Does the first balance have a σ_1 greater than σ_2? Test this hypothesis at the 5% level of significance.

For this case,

$$H_0 : \sigma_1^2 = \sigma_2^2$$

$$H_1 : \sigma_1^2 > \sigma_2^2$$

TABLE 17.4

Critical Regions: F Distribution

Alternative Hypothesis (H_1)	Critical Region
$\sigma_1^2 \neq \sigma_2^2$	$s_1^2/s_2^2 \left\langle F_{1-\alpha/2} \text{ or } s_1^2/s_2^2 \right\rangle F_{\alpha/2}$
$\sigma_1^2 > \sigma_2^2$	$s_1^2/s_2^2 > F_\alpha$
$\sigma_1^2 < \sigma_2^2$	$s_1^2/s_2^2 < F_\alpha$

This is a one-sided (right-tail) test. Noting that $\alpha=0.05$, $v_1=4$, and $v_2=9$, Table 13.5 yields

$$F_{0.05;4,9} = 6.422$$

The critical region for the test is obtained from Table 17.4. The variance ratio is

$$\frac{s_1^2}{s_2^2} = \frac{0.50}{0.10}$$

$$= 5.0$$

Because 5.0 is *not* greater than 6.422, the hypothesis H_0 is *accepted*.

REFERENCES

1. L. Theodore and F. Taylor, *Probability and Statistics*, Theodore Tutorials (originally published by USEPA, RTP, NC), East Williston, NY, 1996.
2. S. Shaefer and L. Theodore, *Probability and Statistics Applications in Environmental Science*, CRC Press/Taylor & Francis Group, Boca Raton, FL, 2007.
3. L. Theodore and R. Dupont, *Environmental Health and Hazard Risk Assessment: Principles and Calculations*, CRC Press/Taylor & Francis Group, Boca Raton, FL, 2013.
4. W. Rosenkrantz, *Probability and Statistics for Science, Engineering, and Finance*, CRC Press/Taylor & Francis Group, Boca Raton, FL, 2009.
5. F. Ricci, personal communication to L. Theodore, Manhattan College, Bronx, NY, 2009.

18 Military and Terrorism

There are 17 applications in this chapter. They key on environmental risk calculations that involve both discrete and continuous probability distributions. Discrete distributions that receive attention include binomial, multinomial, hypergeometric, and Poisson. Continuous distributions include normal, log-normal, exponential, Weibull, chi-square (χ^2), and F. Two of the problems are based on activities at the international level.

The bulk of the applications in this chapter have been adapted from the literature.[1-3]

APPLICATION 18.1

Consider the Picatinny arsenal parallel system shown in Figure 18.1. Determine the reliability of this system employing the information provided in the figure.

Solution

Because this is a parallel system,

$$R_p = 1 - (1 - R_A)(1 - R_B)$$

Employing the information from the figure,

$$R_p = 1 - (1 - 0.92)(1 - 0.95)$$
$$= 1.0 - (0.08)(0.05)$$
$$= 0.9992 = 99.92\%$$
$$R_A = 0.92$$

APPLICATION 18.2

Consider the following sample example. A battery employed at a military facility is deemed reliable if it operates for more than 600 h. The lives of the previous 11 batteries, in hours, were

501, 591, 621, 386, 942, 503, 201, 1013, 902, 32, 899

Solution

Assuming all batteries come from the same population, one notes that 5 of the 11 did not function beyond 500 h. Therefore, the reliability of the battery is simply given by as follows:

$$R = \frac{5}{11}$$
$$= 0.454 = 45.4\%$$

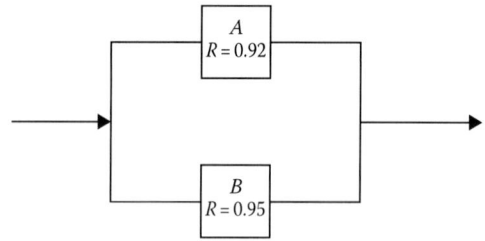

FIGURE 18.1 Parallel system (see Application 18.1).

APPLICATION 18.3

A Central Intelligence Agency (CIA) agent's ability to distinguish a terrorist from a nonterrorist is tested independently on 10 different occasions. What is the probability of 8 correct identifications if the agent is only guessing?

Solution

For guessing purposes, the probability of making a correct terrorist identification may be reasonably assumed to be 0.5. Let X denote the number of correct identifications. If the agent is only guessing, X has a binomial distribution with $n = 10$ and $p = 0.5$. The probability of at least 7 correct identifications is

$$P(X \geq 7) = P(X = 7) + P(X = 8) + P(X = 9) + P(X = 10)$$

The calculations are (noting that $(0.5)^{10} = 0.0009766$)

$$\begin{aligned}
P(X = 7) &= \frac{10!}{(7!)(3!)}(0.5)^7 (0.5)^3 \\
&= (120)(0.0009766) \\
&= 0.1172 = 11.72\%
\end{aligned}$$

$$\begin{aligned}
P(X = 8) &= \frac{10!}{(8!)(2!)}(0.5)^8 (0.5)^2 \\
&= (45)(0.0009766) \\
&= 0.04395 = 4.395\%
\end{aligned}$$

$$\begin{aligned}
P(X = 9) &= \frac{10!}{(9!)(1!)}(0.5)^9 (0.5)^1 \\
&= (10)(0.0009766) \\
&= 0.00976 = 1\%
\end{aligned}$$

$$\begin{aligned}
P(X = 10) &= \frac{10!}{(10!)(0!)}(0.5)^{10} (0.5)^0 \\
&= 0.0009766 = 0.1\%
\end{aligned}$$

Thus,

$$P(X \geq 7) = 0.1172 + 0.04395 + 0.00976 + 0.0009766$$
$$= 0.1719 - 17.2\%$$

APPLICATION 18.4

A drone general hits its target with a probability p of 30%. Find the number of drones that should be fired so that there is at least a 95% probability of hitting the target.

Solution

The probability of missing the target is $q = 1 - p = 0.7$. Hence, the probability that n drones miss the target is $(0.7)^n$. Thus, one seeks the smallest n for which

$$1 - (0.7)^n > 0.95$$

or equivalently,

$$(0.7)^n < 0.05$$

By trial and error,

$$(0.7)^1 = 0.7$$
$$(0.7)^2 = 0.49$$
$$(0.7)^3 = 0.343$$
$$(0.7)^4 = 0.240$$
$$(0.7)^7 = 0.0823$$
$$(0.7)^8 = 0.118$$
$$(0.7)^9 = 0.0823$$
$$(0.7)^{10} = 0.0576$$
$$(0.7)^{11} = 0.0383$$

Thus, at least 11 drones should be fired.

APPLICATION 18.5

If the probability of a control valve developed at a military facility being defective is 10%, what is the probability of obtaining less than 4 defects in a sample size of 4?

Solution

Apply the binomial distribution equation with $p=0.1$, $q=0.90$, $n=4$, and $x=1, 2, 3$.

$$P = P(X=0) + P(X=1) + P(X=2) + P(X=3)$$

In effect,

$$P = \sum_{x=0}^{3} \frac{4!}{x!(4-x)!} (0.1)^x (0.9)^{4-x}$$

The four terms may now be calculated as follows:

$$P(X=0) = (0.9)^4 = 0.6561$$
$$P(X=1) = 4(0.1)^1 (0.9)^3 = 0.2916$$
$$P(X=2) = 6(0.1)^2 (0.9)^2 = 0.0486$$
$$P(X=3) = 4(0.1)^3 (0.9)^1 = 0.0036$$

Thus,

$$P = 0.6561 + 0.2916 + 0.0486 + 0.0036$$
$$= 0.9999 = 99.99\%$$

Is this a reasonable result?

APPLICATION 18.6

An armory's firepower consists of approximately 80% rifles (R), 15% machine guns (M), and 5% cannons (C). If a sample of four pieces of firepower is drawn from the armory. What is the probability that 3 rifles, no machine gun, and 1 cannon will be drawn?

Solution

Apply the multinomial distribution provided in Equation 6.9 (from Chapter 6), with $n=4$:

$$P = \frac{4!}{(3!)(0!)(1!)} (0.8)^3 (0.15)^0 (0.05)^1$$
$$= 0.1024 = 10\%$$

APPLICATION 18.7

A small explosive product has a weight which is normally distributed with a mean of 16 lb. It may be regarded as appreciably under the mean weight and an environmental hazard if it is less than 15.75 lb. If the risk of this is to be less than 0.01 (1%), what is the maximum allowable value of the standard deviation?

Solution

This is a one-sided normal distribution test. For this case,

$$P(X \leq 15.75) = 0.01$$

Normalizing,

$$P\left(\frac{X - \mu}{\sigma} \leq \frac{15.75 - 16}{\sigma}\right) = 0.01$$

$$P\left(Z \leq \frac{15.75 - 16}{\sigma}\right) = 0.01$$

From Table 9.1 (in Chapter 9),

$$-\frac{0.25}{\sigma} = -2.326$$

$$\sigma = 0.1074$$

APPLICATION 18.8

Defective bombs pose a major environmental risk. Ten percent of the bombs produced at a certain arsenal turn out to be defective. Find the probability that in a sample of 10 bombs chosen at random, exactly 2 will be defective.

Solution

Note that this may be solved using either the binomial distribution or Poisson distribution. Apply the Poisson distribution for this application. First note that

$$\lambda = np = (10)(0.1) = 1$$

According to the Poisson distribution,

$$P(X = x) = \frac{\lambda^x e^{-\lambda}}{x!}$$

Substituting,

$$P(X=2) = \frac{(1)^2 e^{-1}}{2!}$$
$$= 0.1839 = 18.4\%$$

Calculating this probability employing the binomial distribution is left as an exercise for the reader.

APPLICATION 18.9

If the probability that 9 soldiers will suffer a bad reaction from handling a poisonous gas is 0.001, determine the probability that out of 2000 soldiers exactly 3 will suffer a bad reaction. Also calculate the probability that more than 2 will experience a bad reaction.

Solution

Let X denote the number of soldiers suffering a bad reaction. Apply the Poisson distribution, that is,

$$P(X=x) = \frac{\lambda^x e^{-\lambda}}{x!}$$

where

$$\lambda = np = (2000)(0.001) = 2$$

Therefore,

$$P(X=3) = \frac{(2)^3 e^{-2}}{3!}$$
$$= 0.180 = 18\%$$

Also,

$$P(X>2) = 1 - \left[P(X=0) + P(X=1) + P(X=2) \right]$$
$$= 1 - \left[\frac{(2)^0 e^{-2}}{0!} + \frac{(2)^1 e^{-2}}{1!} + \frac{(2)^2 e^{-2}}{2!} \right]$$
$$= 1 - 5e^{-2}$$
$$= 0.323 = 32.3\%$$

APPLICATION 18.10

A random sample of 64 observations of rocket travel distance is normally distributed with mean $\bar{X} = 101$ miles and standard deviation 4 miles. Find $P(\bar{X} > 103$ miles$)$.

Solution

If $\bar{X}$ is the mean of a sample from a normal population with mean μ and standard deviation σ, then $\bar{X}$ is normally distributed with mean μ and standard deviation $\sigma/\sqrt{n}$. If the population sampled is not normal, then $\bar{X}$ is approximately normally distributed with mean μ and standard deviation $\sigma/\sqrt{n}$, provided the sample size n is relatively large. This large sample distribution of the sample mean is based on the central limit theorem.[1,2]

For this application, note that $\bar{X}$ is normally distributed with mean 101 and standard deviation $4/\sqrt{64} = 0.5$. Therefore,

$$P(\bar{X} > 102) = P\left(\left[\frac{\bar{X} - 101}{0.5}\right] > \left[\frac{102 - 101}{0.5}\right]\right)$$
$$= P(Z > 2)$$
$$= 0.023 = 2.3\%$$

APPLICATION 18.11

The life expectancy, X, of an Al-Qaeda terrorist has been estimated to be 44 years. If the terrorist's life expectancy is normally distributed with a variance of 16 years, calculate the probability a terrorist will not survive his or her 50th birthday.

Solution

For this application,

$$\sigma = \sqrt{16} = 4$$

This is a one-tailed calculation so that

$$P(X > 50) = P\left(\left[\frac{X - 44}{4}\right] > \left[\frac{50 - 44}{4}\right]\right)$$
$$= P(Z > 1.5)$$

From Table 9.1 (from Chapter 9),

$$P = 0.061 = 6.1\%$$

APPLICATION 18.12

Because of environmental risk concerns, the army is interested in estimating the average distance, in miles, of successful rocket launches. It desires to estimate the true value within 1.0 mile with 95% confidence. An earlier study of 100 rocket launches provided a mean of 88.3 miles with a standard deviation of 8.6 miles. How many additional launches are required to obtain the desired accuracy?

Solution

For this case,

$$P\left(\left[\frac{\bar{X}-88.3}{\sigma/\sqrt{n}}\right] < Z < \left[\frac{\bar{X}+88.3}{\sigma/\sqrt{n}}\right]\right) = 0.95$$

with

$$Z = \frac{\bar{X}+\mu}{\sigma/\sqrt{n}}$$

For the aforementioned equality to apply with 85% confidence,

$$Z = 1.96$$

For a 1 mile difference,

$$\bar{X} - 88.3 = 1.0$$

with

$$Z = 1.96 = \frac{1.0}{8.6/\sqrt{n}}$$

Solving gives

$$\sqrt{n} = \frac{(1.96)(8.6)}{1.0}$$
$$n = 284$$

Therefore, a total of 284 rockets need to be tested. An additional 184, that is, 284 − 100 rockets are required.

APPLICATION 18.13

The lifetime for tanks have a log-normal distribution with $\mu = 4.2$ years and $\sigma = 0.48$ years. Calculate the probability that a tank will last more than 20 years.

Solution

The describing equation becomes

$$P(T > 20) = P(\ln T > \ln 20) = P(\ln T > 3.0)$$

Converting to the standard normal variable,

$$P\left(\ln T > 3.0\right) = P\left[\left(\frac{\ln T - \alpha}{\beta}\right) > \left(\frac{3.0 - \alpha}{\beta}\right)\right]$$

$$= P\left[Z > \left(\frac{3.0 - 4.2}{0.48}\right)\right]$$

$$= P(Z > -2.5)$$

From Table 9.1 (in Chapter 9),

$$P = 1 - 0.006$$
$$= 0.994 = 99.4\%$$

APPLICATION 18.14

The loss of manpower at a fort due to mental illness, Y, has a log-normal distribution. If $\ln Y$ has a mean of 1.5 and variance 1.0, find $P(1 < Y < 10)$.

Solution

If Y has a log-normal distribution, $\ln Y$ has a normal distribution with a mean of 2 and standard deviation $\sigma = 1.0^{1/2} = 1.0$. Therefore,

$$P\left(1 < Y < 10\right) = P(0 < \ln Y < 2.30)$$

Converting to the standard normal variable,

$$P\left(0 < \ln Y < 2.30\right) = P\left[\left(\frac{0 - 1.5}{1}\right) < \left(\frac{\ln Y - 1.5}{1}\right) < \left(\frac{2.30 - 1.5}{1}\right)\right]$$

$$= P\left(-1.5 < Z < +0.80\right)$$

From Table 9.1,

$$P = 0.433 + 0.288$$
$$= 0.721 = 72.1\%$$

APPLICATION 18.15

Consider the two-stage rocket system shown in Figure 18.2. Determine the reliability, R, if the operating time for each unit is approximately 7000 h. Components A and B have exponential failure rates, λ, of 2×10^{-6} and 2.5×10^{-6} failures/h, respectively, where $R_i = e^{-\lambda_i t}$; $t = $ time, h. The term λ may be viewed as the reciprocal of the average time to failure.

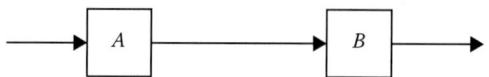

FIGURE 18.2   Exponential failure rate: Series system.

Solution

Because this is a series system,

$$R_s = R_A R_B$$

As indicated earlier, for an exponential failure rate

$$R_i = e^{-\lambda_i t}; \quad t = \text{time, h}$$

so that

$$R_A = e^{-(0.000002)(7000)} = e^{-0.014}$$
$$= 0.986 = 98.6\%$$

and

$$R_B = e^{-(0.0000025)(7000)} = e^{-0.0175}$$
$$= 0.9827 = 98.27\%$$

Therefore,

$$R_s = (0.9860)(0.9827)$$
$$= 0.970 = 97\%$$

APPLICATION 18.16

The entire life of a Jeep has a Weibull distribution with a failure rate of

$$Z(t) = 2t^{-1/2}$$

where t is measured in years. What is the probability that the Jeep will last at least 4 years?

Solution

The pdf specified by $f(t)$ in terms of the failure rate, $Z(t)$, is as follows:

$$f(t) = Z(t) \exp\left(-\int_0^t Z(t)\, dt\right)$$

Substituting $2t^{-1/2}$ for $Z(t)$ yields

$$f(t) = 2t^{-1/2} \exp\left(-\int_o^t 2t^{-1/2}dt\right)$$

$$= 4t^{-1/2} \exp(-2t^{-1/2}); \quad t > 0$$

Employ the integration procedure presented in Illustrative Example 12.2. The probability that the Jeep lasts at least 4 years is

$$P(T \geq 4) = \int_4^\infty 4t^{-1/2} \exp(2t^{-1/2})dt$$

$$= -4e^{-2t^{0.5}} \Big|_4^8$$

$$= -0 - \left(-4e^{-4}\right)$$

$$= 0.0732 = 7.32\%$$

APPLICATION 18.17

The data in Table 18.1 are provided for two methods of measuring for the terrorist activities at the international level. Do the variances produced by the two methods differ significantly? Employ a 2% level of significance.

Solution

F calculated for the two methods is

$$F = \frac{s_1^2}{s_2^2} = \frac{(400)^2}{(201)^2}$$

$$= 4.0$$

At the 2% level of significance with $v_1 = 7$ and $v_2 = 20$, one obtains (because this is a two-sided test) from Table 13.5 (in Chapter 13)

$$F_{0.01;7,20} = 3.7 \left(\text{by interpolation}\right)$$

Because $3.7 < 4.0$, the difference between the variances *is* significant.

TABLE 18.1
Terrorist Activities

Method	Sample Size	Standard Deviation(ppm)
1	8	400
2	21	201

APPLICATION 18.18

A sample of 300 explosives from 1 day's production of an arsenal pilot plant (where the author once worked) is classified by work shift and quality as follows:

Quality	Day	Evening	Midnight	Total
Poor	12	12	10	34
Satisfactory	90	68	72	230
Excellent	21	10	5	36
Total	123	90	87	300

Test the assumption that quality of explosives produced is independent of the work shift in which they are produced. Use a 10% level of significance.[2]

Solution

The product of the row total and the column total divided by the grand total is computed for each cell as follows:

$$e_{11} = \frac{(34)(123)}{300} = 13.9$$

$$e_{12} = \frac{(34)(90)}{300} = 10.2$$

$$e_{13} = \frac{(34)(87)}{300} = 9.9$$

$$e_{21} = \frac{(230)(123)}{300} = 94.3$$

$$e_{22} = \frac{(230)(90)}{300} = 69.0$$

$$e_{23} = \frac{(230)(87)}{300} = 66.7$$

$$e_{31} = \frac{(36)(123)}{300} = 14.8$$

$$e_{32} = \frac{(36)(90)}{300} = 10.8$$

$$e_{33} = \frac{(36)(87)}{300} = 10.4$$

Note that e_{ij} represents the theoretical frequency for the cell at the intersection of the ith row and the jth column.

The test statistic is given by

$$\sum_{i=1}^{r}\sum_{j=1}^{c}\frac{(f_{ij}-e_{ij})^2}{e_{ij}}$$

where
    f_{ij} is the observed frequency of the ith row and jth column, that is, the cell at the intersection of the ith row and the jth column
    e_{ij} is the corresponding theoretical frequency
    r is the number of rows
    c is the number of columns

Therefore, the chi-square test statistic is

$$\chi^2 = \frac{(12-13.9)^2}{13.9} + \frac{(12-10.2)^2}{10.2} + \frac{(10-9.9)^2}{9.9} + \frac{(90-94.3)^2}{94.3} + \frac{(68-69)^2}{69}$$
$$+ \frac{(72-66.7)^2}{66.7} + \frac{(21-14.8)^2}{14.8} + \frac{(10-10.8)^2}{10.8} + \frac{(5-10.4)^2}{10.4}$$
$$= 6.67$$

The probability (P-value) that a chi-square variable with $(r-1)(c-1)$ degrees of freedom exceeds the value of the chi-square test statistic from Table 13.5 is

$$P\left(\chi^2_{(r-1)(c-1)} > 6.67\right) > 0.10; r = 3, c = 3, (r-1)(c-1) = (2)(2) = 4$$
$$P\left(\chi^2_4 > 6.67\right) > 0.10$$

Because the P-value exceeds 0.10, one would fail to reject the assumption that the quality of an explosive is independent of the work shift during which it was produced. Note that rejection of independence implies that the row and column criteria of classification are interrelated.

    The reader may choose to compare this Illustrative Example with that presented in Chapter 16 (Illustrative Example 16.7).

REFERENCES

1. L. Theodore and F. Taylor, *Probability and Statistics*, Theodore Tutorials (originally published by USEPA, RTP, NC), East Williston, NY, 1996.
2. S. Shaefer and L. Theodore, *Probability and Statistics Applications in Environmental Science*, CRC Press/Taylor & Francis Group, Boca Raton, FL, 2007.
3. L. Theodore and R. Dupont, *Environmental Health and Hazard Risk Assessment: Principles and Calculations*, CRC Press/Taylor & Francis Group, Boca Raton, FL, 2013.

19 Travel/Aerospace/ Weather

There are 19 applications in this chapter. They key on environmental risk calculations that involve both discrete and continuous probability distributions. Discrete
distributions that receive attention include binomial, multinomial, hypergeometric,
and Poisson. Continuous distributions include normal, log-normal, exponential,
Weibull, chi-square (χ^2), F, and one joint probability distribution function (JPDF).
Two of the applications are based on activities at the international level.

The bulk of the applications have been adapted from the literature.[1–3]

APPLICATION 19.1

Let X denote the number of trucks reaching a waste site on a given day. The pdf for
X is given by

$$f(x) = \frac{x}{36}; \quad x = 1,2,3,4,5,6,7,8$$

1. Verify that $f(x)$ is a valid pdf, and
2. Calculate the probability that less than 6 trucks will arrive on a given day.

Solution

1. For $f(x)$ to be valid, the following two conditions must be satisfied:

$$0 \le f(x) \le 1; \quad x \text{ between 1 and 6}$$

and

$$\sum_{1}^{6} f(x) = 1$$

Substituting into the second condition yields

$$\sum_{1}^{6} f(x) = \sum_{1}^{6} \left(\frac{x}{36} \right)$$

$$= \frac{1}{36} + \frac{2}{36} + \frac{3}{36} + \frac{4}{36} + \frac{5}{36} + \frac{6}{36} + \frac{7}{36} + \frac{8}{36}$$

$$= \frac{36}{36} = 1.0$$

Thus, both the conditions are satisfied and $f(x)$ is a valid pdf.

2. The probability that at least 5 trucks will arrive on a given day can be determined by

$$P(0 \leq X < 6) = f(x=0) + f(x=1) + f(x=2) + f(x=3) + f(x=4) + f(x=5)$$
$$= \frac{0}{36} + \frac{1}{36} + \frac{2}{36} + \frac{3}{36} + \frac{4}{36} + \frac{5}{36}$$
$$= \frac{15}{36}$$
$$= 0.417 = 41.7\%$$

APPLICATION 19.2

It would be appropriate to define the year 2010 as the year of natural disasters. The natural disasters in this case were primarily earthquakes, with three major earthquakes rocking the planet that year.

The January 2010 Haiti earthquake had a Richter scale magnitude of 7.0. The Haitian government reported that an estimated 230,000 people had died, 300,000 had been injured, and 1,000,000 had been made homeless as a result of the quake. They also estimated that 250,000 residences and 30,000 commercial buildings had either collapsed or were severely damaged. It was the most severe earthquake to hit Haiti in 200 years. It should also be noted that construction standards are poor in Haiti; the country has no building codes. At the time of the preparation of this chapter, as much as 98% of the rubble from the quake remained. An estimated 26 million cubic yards remain, making most of the capital impassable. Thousands of bodies remain in the rubble. The number of people in relief camps is estimated at 1.6 million, and almost no transitional housing has been built. Most of the camps have no electricity, running water, or sewage disposal, and tents are beginning to fall apart. Crime in the camps is widespread, especially against women and children. The O'Reilly Factor (FOX News) has reported that little to none of the relief money has been dispersed to its intended beneficiaries.[3]

One of the largest earthquakes in recorded history destroyed houses, bridges, and highways in central Chile in late February 2010 and sent a tsunami rocking halfway around the globe. Chileans near the epicenter were tossed around as if shaken by a giant. The 8.8 magnitude quake was felt as far away as Sao Paulo in Brazil—nearly 2000 miles away. The full extent of the damage still remains unclear at that time as dozens of aftershocks shuddered across the nation.[3]

The China earthquake of 2010 registered 7.1 on the Richter scale and resulted in the deaths of hundreds of thousands. In Jiegu, a township near the epicenter, more than 85% of houses collapsed, while large cracks appeared in buildings still standing. Officials initially reported that 10,000 people were injured due to the quakes that hit Yushu County in Qinghai Province. It was concluded that people died because of the collapse of cheaply constructed buildings in a poor region where it seems little regard had been paid to building codes that could have offered better protection to the people.[3]

Based on the aforementioned facts, Theodore and Dupont Associates (TADA) estimated that the difference between that magnitude of a large earthquake, as measured
on the Richter scale, and the threshold value of 3.25 at a South American location is
a random variable X having the pdf

$$f(x) = 1.7e^{-1.7x}; \quad x > 0$$
$$= 0; \quad \text{elsewhere}$$

Calculate $P(2 < X < 6)$.

Solution

Applying the definition of cdf leads to

$$P(2 < X < 6) = \int_{2}^{6} f(x) dx$$

Substituting and integrating yields

$$P(2 < X < 6) = \int_{2}^{6} 1.7e^{-1.7x} dx$$
$$= e^{-3.4} - e^{-10.2}$$
$$= 0.0334 - 0.0$$
$$= 0.0334 = 3.34\%$$

APPLICATION 19.3

Let X denote the annual number of hurricanes in a certain region. The pdf of X is
specified as:

$$f(x) = 0.25; \quad x = 0$$
$$f(x) = 0.35; \quad x = 1$$
$$f(x) = 0.24; \quad x = 2$$
$$f(x) = 0.11; \quad x = 3$$
$$f(x) = 0.04; \quad x = 4$$
$$f(x) = 0.01; \quad x = 5$$

What is the probability of at least 4 hurricanes in any year?

Solution

For at least 4 hurricanes in any year, one may write

$$P(x \geq 4) = \sum_4^\infty f(x) = \sum_4^5 f(x)$$
$$= f(4) + f(5)$$
$$= 0.04 + 0.01$$
$$= 0.05 = 5\%$$

APPLICATION 19.4

Mine safety began with the founding of the Bureau of Mines in 1910. Mine accidents were defined at that time as those resulting in five or more deaths. It was the Monongah coal mine explosion in 1907, claiming 362 lives, that prompted Congress to create the Bureau of Mines in 1910 (Public Law 61-179). Mine accidents have declined dramatically in number and severity since the Bureau's inception, with mine accidents resulting in five or more deaths becoming a rare occurrence in modern mining operations.[3]

The most recent accident (just before the initial preparation of this chapter) that received international attention occurred in Chile. The drama of the rescue of 33 miners trapped in a collapsed mine was viewed on TV by nearly the entire world, and it was clearly demonstrated that both the health and hazard risks miners are exposed to on a daily basis. On a human interest level, Chilean miner Edison Pena carried a Chilean flag to the finish line of the New York City marathon only 3½ weeks after being rescued from his 69 day ordeal at the bottom of the San Jose mine.[3]

The data in Table 19.1 on mining accidents in Chile were recently published by Sernageomin, the National Geologic and Mineral Service of Chile. The mining industry in Chile currently involves 42 mine processing sites and a total of 175,000 mine workers. Based on the data given in Table 19.1 for the last 10 years of accident statistics, determine the following:

1. Calculate the annual risk of mining injuries in units of injuries/miner, and injuries/million man-hours worked.
2. Determine the annual risk of mining deaths in units of deaths/miner, and deaths/million man-hours worked.

Solution

The average accident rate over the last 10 years in the Chilean mining industry expressed as average accident/year, average deaths/year, and average man-hours worked/year is as follows: 30 accidents/year, 1789 injuries/year, 35 deaths/year, and 283 million man-hours worked/year.

TABLE 19.1
Historical Chilean Mining Data, 1997–2010

Year	Number of Mining Accidents	Number of Mining Injuries	Number of Mining Deaths	Million Man-Hours Worked
1997	43	3211	49	227
1998	33	2248	38	239
1999	17	1607	19	203
2000	32	1452	37	205
2001	30	1917	34	227
2002	25	2030	28	240
2003	26	1767	30	233
2004	29	1877	33	256
2005	26	1869	30	285
2006	27	1679	31	284
2007	35	1872	40	321
2008	38	2021	43	361
2009	31	1391	35	354
2010	36	1800	40	350

Average annual risk (AAR) of a mining-related injury:

$$\left(AAR\right)\text{-related injury} = \frac{1{,}789 \text{ injuries/year}}{175{,}000 \text{ miners/year}}$$

$$= 0.01 \text{ injuries/miner}$$

$$\left(AAR\right)\text{-related injury} = \frac{1{,}789 \text{ injuries/year}}{283 \text{ million man-hours worked/year}}$$

$$= 6.25 \text{ injuries/million man-hours worked}$$

$$\left(AAR\right)\text{-related deaths} = \frac{35 \text{ deaths/year}}{175{,}000 \text{ miners/year}}$$

$$= 0.0002 \text{ deaths/miner}$$

$$\left(AAR\right)\text{-related deaths} = \frac{35 \text{ deaths/year}}{283 \text{ million man-hours worked/year}}$$

$$= 0.12 \text{ deaths/million man-hours worked}$$

APPLICATION 19.5

The probability is 0.80 that an airplane crash due to structural failure is diagnosed correctly. Suppose, in addition, that the probability is 0.30 that an airplane crash not due to structural failure is incorrectly attributed to structural failure. If 35% of all airplane crashes are due to structural failure, what is the probability that an airplane crash was due to structural failure, given that it has been so diagnosed?

Solution

Let A_1 be the event that structural failure is the cause of the airplane crash. Let A_2 be the event that the cause is other than structural failure. Let B be the event that the airplane crash is diagnosed as being due to structural failure. Then

$$P(A_1) = 0.35; \quad P(B \mid A_1) = 0.80$$
$$P(A_2) = 0.65; \quad P(B \mid A_2) = 0.30$$

Substituting in Equation 3.36 (in Chapter 3), one obtains

$$P(A_1 \mid B) = \frac{P(A_1)P(B \mid A_1)}{P(A_1)P(B \mid A_1) + P(A_2)P(B \mid A_2)}$$

$$P(A_1 \mid B) = \frac{(0.35)(0.80)}{(0.35)(0.80) + (0.65)(0.30)} = 0.59$$

Therefore, the diagnosis of structural failure revises its probability as the cause of the airplane crash upward from 0.35 to 0.59.

APPLICATION 19.6

A failure analysis of a military overseas flight is regarded as a series system with the following components: ground crew (A), cockpit crew (B), aircraft (C), weather conditions (D), and landing accommodations (E). The cockpit crew is viewed as a parallel system with the following components: captain (B_1), copilot (B_2), and flight engineer (B_3). Landing accommodations are viewed as a parallel system with the following components: scheduled airport (E_1) and alternate landing sites (E_2 and E_3). Failure probabilities for the various components are estimated as follows:

$A = 0.001$	$B_3 = 0.100$	$E_1 = 0.001$
$B_1 = 0.001$	$C = 0.001$	$E_2 = 0.050$
$B_2 = 0.010$	$D = 0.0001$	$E_3 = 0.100$

What is the probability of a successful flight?

Solution

First identify the components connected in parallel. B_1, B_2, and B_3 are connected in parallel. E_1, E_2, and E_3 are also connected in parallel. First, compute the reliability of each subsystem of the components connected in parallel. The reliability of a parallel subsystem consisting of components B_1, B_2, and B_3 is

$$R_p = 1 - (1 - 0.999)(1 - 0.99)(1 - 0.90) = 0.999999$$

The reliability of the parallel subsystem consisting of the components E_1, E_2, and E_3 is

$$R_p = 1 - (1 - 0.999)(1 - 0.95)(1 - 0.90) = 0.999995$$

One may now multiply the product of the reliabilities of the parallel subsystems by the product of the reliabilities of the components to which the parallel subsystems are connected in series:

$$R_s = (0.999999)(0.999995)(0.999)(0.999)(0.9999) = 0.9979$$

The probability of a successful flight is therefore 0.9979 or 99.79%. Would the reader consider going away without leave (AWOL) if assigned to this flight?

APPLICATION 19.7

A procuring agent for National Aeronautics and Space Administration (NASA) is asked to sample a lot of 100 pumps for possible employment in a spacecraft. The sample procedure calls for the inspection of 20 pumps. If there are any bad pumps, the lot is rejected; otherwise, it is accepted. The chief engineer has asked the agent the following question. Suppose at the most 2 bad pumps are allowed in the sample. What kind of protection would NASA have?

Solution

For this case, one needs to calculate the probability of accepting the lot (with 4 defective pumps) if 0, 1, or 2 defectives is allowed in the sample. Therefore,

$$P(X \le 2) = P(X = 0) + P(X = 1) + P(X = 2)$$

$$= \frac{20!}{(0!)(20!)}(0.04)^0 (0.96)^{20} + \frac{20!}{(1!)(19!)}(0.04)^1 (0.96)^{19}$$

$$+ \frac{20!}{(2!)(18!)}(0.04)^2 (0.96)^{18}$$

$$= 0.442 + (20)(0.04)(0.4604) + (10)(19)(0.0016)(0.48)$$

$$= 0.442 + 0.368 + 0.146$$

$$= 0.956 = 95.6\%$$

APPLICATION 19.8

A Pentagon quality control engineer inspects a random sample of 3 explosives drawn *without replacement* from each incoming group of 100 explosives. An entire group is accepted only if the sample contains a certain (minimum) number of defectives. What is the probability of accepting a group containing 10 defective explosives? Generate the equation describing this application.

Solution

The solution requires the application of the hypergeometric distribution equation. Each group consists of 50 explosives. Therefore, $n = 100$. For this case, associate "success" with a defective explosive. Determine a, the number of "successes" in the set of n items. For this problem, $a = 10$. A sample of four explosives is drawn without replacement from each lot; therefore, $r = 3$.

Substitute the values of n, r, and a in the pdf of the hypergeometric distribution, that is, Equation 7.1 (in Chapter 7):

$$f(x) = \frac{\dfrac{10!}{x!(10-x)!} \dfrac{97!}{(3-x)!(94+x)!}}{\dfrac{100!}{(3!)(94!)}}; \quad x = 0,1,2,3,4$$

APPLICATION 19.9

The number of hazardous waste trucks arriving daily at a certain hazardous waste incineration facility has a Poisson distribution with parameter n. Present facilities can accommodate 3 trucks a day. If more than 3 trucks arrive in a day, the trucks in excess of 3 must be sent elsewhere. How much must the present waste facilities be increased to permit handling of all trucks for 95% of the days?

Solution

Note that the number of trucks has not been specified. The general solution is given as follows:

$$P(\text{sending trucks elsewhere}) = 0.05 = 1 - \sum_{i=0}^{n} P(X = i)$$

A trial-and-error solution is required because a different equation is obtained for each value of n. For example, if $n = 5$,

$$0.05 = 1 - \left[P(X = 0) + P(X = 1) + P(X = 2) + P(X = 3) + P(X = 4) + P(X = 5) \right]$$

Substituting gives

$$0.05 = 1 - e^{-2.5}\left[1 + 2.5 + \frac{(2.5)^2}{2} + \frac{(2.5)^3}{6} + \frac{(2.5)^4}{24} + \frac{(2.5)^5}{120}\right]$$

$$= 1 - 0.082\left[1.0 + 2.5 + 3.125 + 2.604 + 1.628 + 0.814\right]$$

$$= 1 - 0.958$$

$$0.05 \approx 0.042$$

The increase is from 3 to 5; therefore, the facilities must be increased by approximately 67%.

APPLICATION 19.10

The mean weight of an astronaut is 151 lb and the standard deviation is 15 lb. Assuming that the weights are normally distributed, find how many astronauts weigh

1. Between 120 and 155 lb, and
2. More than 185 lb

Solution

First note that the weight of the astronauts recorded as being between 120 and 155 lb can actually have any value from 119.5 to 155.5 lb.

1. Assuming this to be the case, calculate the standard normal variable for the two weights.

$$Z_1(119.5 \text{ lb}) = (119.5 - 151)/15$$
$$= -2.10$$
$$Z_2(155.5 \text{ lb}) = (155.5 - 151)/15$$
$$= 0.30$$

From Table 9.1 (in Chapter 9),

$$P(-2.10 < Z < +0.30) = 0.4821 + 0.1179$$
$$= 0.6000 = 60\%$$

The number of astronauts weighing between 120 and 155 lb is, therefore,

$$500(0.6000) = 300$$

2. Assume now that astronauts weighing more than 185 lb must weigh at least 185.5 lb. Calculate Z.

$$Z(185.5\,\text{lb}) = (185.5 - 151)/15$$
$$= 2.30$$
$$P(Z > 2.30) = 0.5 - 0.4893$$
$$= 0.0107 = 1.1\%$$

The number of astronauts weighing more than 185 lb is therefore approximately

$$500(0.0107) = 5$$

APPLICATION 19.11

The normalized wind speeds at a certain hurricane test site during the past 10 years are summarized in Table 19.2. If the wind speeds are assumed to follow a *log-normal* distribution, predict the level that would be exceeded once in 20 years.

Solution

For this case, refer to Table 19.2. Based on the data and calculations presented in Table 19.3 (see also details in Chapter 10),

$$\bar{X} = 3.886$$
$$s^2 = 0.64$$

TABLE 19.2
Wind Speed Data

Year	1	2	3	4	5	6	7	8	9	10
Wind speed, u, (ft/min)	23	38	17	210	62	142	43	29	71	31

TABLE 19.3
Wind Speed Calculations

Year (Y)	Wind Speed, u	$X = \ln u$	X^2
1	23	3.13	9.83
2	38	3.64	13.25
3	17	2.83	8.01
4	210	5.35	28.62
5	62	4.13	17.06
6	142	4.96	24.60
7	43	3.76	14.14
8	29	3.37	11.36
9	71	4.26	18.15
10	31	3.43	11.76
Total	—	38.86	156.78

and

$$s = 0.80$$

For this test, with $Z = 1.645$ for 95% (once in 20 years, that is, $1/20 = 0.05$) value,

$$Z = \frac{X - \bar{X}}{s}$$

$$1.645 = \frac{X - 3.886}{0.80}$$

Solving for X yields

$$X = 5.202$$

One may conclude for this log-normal distribution, that

$$X = 5.202 = \ln(u)$$

so that

$$u = 259 \text{ ft/min}$$

APPLICATION 19.12

The time to failure for an air conditioner in a spaceship is presumed to follow an exponential distribution with $\lambda = 0.1$ (per year). What is the probability of a failure within the first 3 years?

Solution

Refer to Chapter 11.

$$F(x) = \int_0^x \lambda e^{-\lambda x}\, dx = 1 - e^{-\lambda x}$$

For this case,

$$P(X \le 3) = \int_0^3 (0.1) e^{-(0.1)x}\, dx$$

$$= -\frac{0.1}{0.1} e^{-(0.1)x}\Big|_0^3$$

$$= e^{-0.3} + e^0$$

$$= 1 - e^{-0.3} = 1 - 0.741$$

$$= 0.259 = 25.9\%$$

Therefore, there is nearly a 26% probability that the air conditioner in the spacecraft will fail within the first 3 years.

APPLICATION 19.13

The probability that an anemometer at a weather station will not survive for more than 144 months is 0.99. How often should the anemometer be replaced?

Solution

This requires the calculation of μ in the exponential model with units of (month)$^{-1}$. Once again

$$F(t) = 1 - e^{-\lambda t}$$

Based on the information provided,

$$P(T \le 144) = 0.99$$

or

$$0.99 = 1 - e^{-(\lambda)(144)}$$

Solving for λ gives

$$\lambda = -\frac{1}{144} \ln(1 - 0.99)$$
$$= 0.032$$

Because the expected time (or life), $E(T)$, is

$$E(T) = \frac{1}{0.032}$$
$$= 31.3 \text{ months}$$

The anemometer should therefore be replaced in approximately 2½ years.

APPLICATION 19.14

The entire life of a foreign-made car has a Weibull distribution with failure rate

$$Z(t) = 2.15t^{-1/2}$$

where t is measured in years. What is the probability that the car will last at least 4 years?

Solution

The pdf specified by $f(t)$ in terms of the failure rate, $Z(t)$, is as follows:

$$f(t) = Z(t)\exp\left(-\int_0^t Z(t)\,dt\right)$$

Substituting $2.15t^{-1/2}$ for $Z(t)$ yields

$$f(t) = 2.15t^{-1/2}\exp\left(-\int_0^t 2.15t^{-1/2}\,dt\right)$$

$$= (2.15)^2 t^{-1/2}\exp(-2.15t^{-1/2}); \quad t > 0$$

$$= 4.62t^{-1/2}\exp(-2.15t^{-1/2})$$

Employ the integration procedure described earlier. The probability that the car lasts at least 4 years is

$$P(T \geq 4) = 4.62\int_4^\infty t^{-1/2}\exp(2.15t^{-1/2})\,dt$$

$$= -4.62e^{-2t^{0.5}}\Big|_4^\infty$$

$$= 4.62\left[-0 - \left(-4e^{-4}\right)\right]$$

$$= 0.0742 = 7.42\%$$

APPLICATION 19.15

Some probability calculations involve the exponential and Weibull distributions introduced in conjunction with the bathtub curve of failure rate. Consider the case of a spaceship transistor having a constant rate of failure of 0.02 per thousand hours. Find the probability that the transistor will operate for at least 25,000 h.

Solution

Substitute the failure rate

$$Z(t) = 0.02$$

into the Weibull equation in Chapter 12, which yields

$$f(t) = \exp\left[-\int_0^t 0.02\,dt\right]$$

$$= 0.02e^{-0.02t}; \quad t > 0$$

as the pdf of T, the time to failure of the transistor. Because t is measured in thousands of hours, the probability that the transistor will operate for at least 25,000 h is given by

$$P(T > 25) = \int_{25}^{\infty} -0.02e^{-0.02t} \, dt$$
$$= -e^{-\infty} + e^{-0.02(25)}$$
$$= 0 + 0.607$$
$$= 0.607 = 61\%$$

APPLICATION 19.16

Calculate the probability that 14 anemometer readings (m/s) from a population with variance of approximately 1.0 will have a sample variance greater than 1.9.

Solution

Assume the population velocity readings are normally distributed. As noted in Equation 13.6 (in Chapter 13), the random variable

$$\frac{(n-1)s^2}{\sigma^2}$$

has a chi-square distributed with $n - 1$ or degrees of freedom. The describing equation is written as

$$P(s^2 > 1.9) = \left(\frac{(n-1)s^2}{\sigma^2} > \frac{14(1.9)}{1.0} \right)$$
$$= (\chi^2 > 26.6)$$

From Table 13.3 (in Chapter 13), with $v = 13$,

$$P(s^2 > 1.9) \cong 0.024 \text{ (linear interpolation)}$$

APPLICATION 19.17

The distance traveled on a gallon of gasoline was measured for a random sample of 16 cars of a particular make and model. The sample mean was 28 miles, and the sample standard deviation was 2.6 miles. As part of an energy conservation study, obtain a 95% confidence interval for the population mean gas mileage, μ, under the assumption that the sample comes from a normal population.

Solution

Note that the sample size is small. Select a statistic involving the sample mean and population mean and having a known pdf under the assumption that $(\bar{X} - \mu)/(s/\sqrt{n})$ has a Student's distribution with $(n-1)$ degrees of freedom. Once again, construct an inequality concerning the statistic selected, such that the probability that the inequality is true equals the desired confidence coefficient. From the table of Student's t distribution,

$$P(-2.131 < v_r < 2.131) = 0.95; \; r = n - 1$$
$$P(-2.131 < v_{n-1} < 2.131) = 0.95$$
$$P(-2.131 < v_{15} < 2.131) = 0.95$$

Therefore,

$$P\left(-2.131 < \frac{\bar{X} - \mu}{s/\sqrt{n}} < 2.131\right) = 0.95$$

Solve the preceding equation for the population mean μ:

$$P\left(-2.131 < \frac{\bar{X} - \mu}{s/\sqrt{n}} < 2.131\right)$$
$$-2.131(s/\sqrt{n}) < \bar{X} - \mu < 2.131(s/\sqrt{n})$$
$$-\bar{X} - 2.131(s/\sqrt{n}) < -\mu < -\bar{X} + 2.131(s/\sqrt{n})$$

Multiplication by -1 requires reversal of the inequality signs:

$$\bar{X} + 2.131(s/\sqrt{n}) > \mu > \bar{X} - 2.131(s/\sqrt{n})$$

or

$$\bar{X} - 2.131(s/\sqrt{n}) < \mu < \bar{X} + 2.131(s/\sqrt{n})$$

Select a basis of 1 gal. Substitute the observed values from the sample to obtain the 95% confidence interval for μ. Note that $\bar{X} = 28$, $n = 16$, and $s = 2.6$.

$$28 - 2.131(2.6/\sqrt{16}) < \mu < 28 + 2.131(2.6/\sqrt{16})$$
$$26.61 < \mu < 29.39$$

Therefore, $26.61 < \mu < 29.39$ is the 95% confidence interval for μ.

APPLICATION 19.18

Two brands of spacecraft batteries are to be compared with respect to variability of battery life. For this purpose, a random sample of eight batteries of Brand I and a random sample of 10 batteries of Brand II are compared with the results in Table 19.4.

Do the brands differ significantly with respect to variability? Assume the populations samples are normal and independent, and that the tolerated probability of a Type I error is 0.02.[1,2]

Solution

First identify the null hypothesis, H_0, and the alternative hypothesis, H_1 as follows:

$$H_0 : \sigma_1^2 = \sigma_2^2$$
$$H_1 : \sigma_1^2 \neq \sigma_2^2$$

where
 σ_1^2 is the population variance of Brand I
 σ_2^2 is the population variance of Brand II

Because the alternative hypothesis is $H_1 : \sigma_1^2 \neq \sigma_2^2$ and $\alpha = 0.02$ (a two-sided test), the critical region is defined by

$$\frac{s_1^2}{s_2^2} < F_{0.99} \quad \text{or} \quad \frac{s_1^2}{s_2^2} > F_{0.01}$$

with

$$n_1 - 1 = 7$$
$$n_2 - 1 = 9$$

For a random variable having an F distribution with 7 and 9 degrees of freedom, $F_{0.01} = 5.61$, that is, the value exceeded with probability of 0.01 by a random variable having an F distribution with 7 and 9 degrees of freedom. As described earlier, $F_{0.99}$ is obtained by making use of the fact that a random variable having an F distribution with 7 and 9 degrees of freedom is the reciprocal of a random variable having an F

TABLE 19.4
Battery Life Data

	Brand I	Brand II
Sample size	$n_1 = 8$	$n_2 = 10$
Sample mean	$\overline{X}_1 = 5$	$\overline{X}_2 = 5$
Sample variance	$s_1 = 2$	$s_2 = 1.2$

distribution with 9 and 7 degrees of freedom. Let $F_{7,9}$ and $F_{9,7}$ represent the original and reciprocal random variables, respectively. One then obtains

$$F_{0.99;7,9} = \frac{1}{F_{0.01;9,7}}$$

$$= \frac{1}{6.72} = 0.15$$

The critical region, therefore, consists of those values of the test statistic, s_1^2/s_2^2, exceeding 5.61 or less than 0.15. The test statistic may now be calculated as follows:

$$\frac{s_1^2}{s_2^2} = \frac{2}{1.2}$$

$$= 1.67$$

Because 1.67 lies between 0.15 and 5.61, H_0 is not rejected. Therefore, Brands I and II do not differ significantly with respect to variability of spacecraft battery life.

APPLICATION 19.19

A *meteor* is defined as a small solid body entering the Earth's atmosphere from outer space. A meteor that reaches the Earth's surface before it is completely consumed is defined as a *meteorite*. Some refer to large meteors/meteorites as *asteroids*, while others refer to asteroids as small planets. Meteorites can be large or small. Most are produced by impacts of larger asteroids. When meteorites enter the atmosphere, frictional forces cause the body to heat up and emit light, thus forming a fireball, also known as a *shooting star* or *falling star*. Finally, meteorites are almost always named for the place where they land, for example, Tunguska (Siberia).

Most meteors disintegrate on entering the atmosphere and typically arrive at the surface at their terminal settling velocity.[4] The impact normally creates a small pit. However, falling meteorites have been known to cause damage to property, animals, and people. Few meteorites are large enough to create large impact craters.

A large meteorite (hereafter referred to as an asteroid) is headed precariously close to the Earth's orbit at an unprecedented velocity. Should an unforeseen force, for example, a collision with a smaller asteroid, cause the asteroid to shift off course, the certainty of where it will cross the plane of the Earth's orbit decreases radically from the predicted point of intersection.

Because of the imminence of this event, the Russians have hired Ricci Associates (RA) to develop a joint probability distribution function (JPDF) that describes the likelihood of where the asteroid will cross the plane of the Earth's orbit. Based on the current flight path, the asteroid will cross the plane of Earth's orbit at approximately 1×10^6 miles from the Earth's effective gravitation influence (EGI). What is the probability that the asteroid's flight will deviate such that it *enters* the Earth's EGI? Assume that the asteroid is moving very fast relative to Earth's orbit.[5]

Solution

Although the details of the solution to this space/travel application have been provided by Ricci,[5] only the final result is presented because of the mathematical complexities involved. The probability of the occurrence of an asteroid collision with the Earth is approximately 2×10^{-23}. Hence, there is (fortunately) an exceedingly small probability that the asteroid will enter the Earth's sphere of EGI based on the JPDF and assumptions set forth by Ricci.[5]

REFERENCES

1. L. Theodore and F. Taylor, *Probability and Statistics*, Theodore Tutorials (originally published by USEPA, RTP, NC), East Williston, NY, 1996.
2. S. Shaefer and L. Theodore, *Probability and Statistics Applications in Environmental Science*, CRC Press/Taylor & Francis Group, Boca Raton, FL, 2007.
3. L. Theodore and R. Dupont, *Environmental Health and Hazard Risk Assessment: Principles and Calculations*, CRC Press/Taylor & Francis Group, Boca Raton, FL, 2013.
4. L. Theodore, *Air Pollution Control Equipment Calculations*, John Wiley & Sons, Hoboken, NJ, 2009.
5. F. Ricci, personal communication, Princeton University, Princeton, NJ, 2011.

20 Nanotechnology

There are 19 illustrative examples in this final chapter. They key on environmental risk calculations that involve both discrete and continuous probability distributions. Discrete distributions that receive attention include binomial, multinomial, hypergeometric, and Poisson. Continuous distributions include normal, log-normal, exponential, Weibull, chi-square (χ^2), F, and a combined health/hazard risk assessment. Three of the problems are based on activities at the international level.

The bulk of the applications have been adapted from the literature.[1–3]

APPLICATION 20.1

The following are ambient particulate concentrations ($\mu g/m^3$) during the first 4 months of operation in a nanolaboratory.

$$22 \quad 17 \quad 6 \quad 10 \quad 5 \quad 13 \quad 12 \quad 16 \quad 21$$

Calculate the following quantities[2]:

1. Mean
2. Median
3. Mode
4. Range
5. Standard deviation
6. Variance

Solution

1. The mean

$$\bar{X} = \frac{\sum x_i}{n}$$
$$= \frac{22 + 17 + 6 + 10 + 5 + 13 + 12 + 16 + 21}{9}$$
$$= 13.55 \ \mu g/m^3$$

2. The middle value is 13 mg/m³.

$$5 \quad 6 \quad 10 \quad 12 \quad \textcircled{13} \quad 16 \quad 17 \quad 21 \quad 22$$

3. There is no mode in this group.
4. Range = maximum value − minimum value

$$= 22 - 5 = 17 \ mg/m^3$$

5. The standard deviation

$$s = \sqrt{\frac{\sum x_i^2 - \dfrac{\left(\sum x_i\right)^2}{n}}{n-1}}$$

Therefore,

$$s = \sqrt{\frac{1944 - \dfrac{14884}{9}}{8}}$$

$$= \sqrt{\frac{290.23}{8}}$$

$$= \sqrt{36.27}$$

$$= 6.02 \text{ mg/m}^3$$

6. Variance = (standard deviation)²

$$s^2 = \left(6.02\right)^2 = 36.27 \ (\text{mg/m}^3)^2$$

APPLICATION 20.2

Exposure to a nanoagent has 4 ways in which it can lead to major health effects (HE) and 12 ways to minor effects. How many ways may an individual be exposed to the nanogent based on the following?

1. 2 major and 2 minor HE; and
2. 1 major and 4 minor HE

Solution

Because the order in which the effects appear or impact an individual does not matter, an analysis involving *combinations without replacement* is required. The number of combines of HE is as follows

1. For 2 major and 2 minor effects

$$HE = C\left(4,2\right)C\left(12,2\right)$$

$$= \left(\frac{4!}{2!2!}\right)\left(\frac{12!}{10!2!}\right)$$

$$= \left(\frac{4 \times 3}{2}\right)\left(\frac{12 \times 11}{2}\right)$$

$$= 396$$

2. For 1 major and 4 minor effects

$$HE = C(4,1)C(12,4)$$
$$= \left(\frac{4!}{3!1!}\right)\left(\frac{12!}{8!4!}\right)$$
$$= (4)\left(\frac{12 \times 11 \times 10 \times 9}{4 \times 3 \times 2}\right)$$
$$= 1980$$

APPLICATION 20.3

An environmental engineering intern submitted the following to describe a continuous random variable, X, at a nanofacility with a pdf given as follows:

$$f(x) = x^2 + kx + 3; \quad 0 \le x \le 4$$

Determine the constant k such that $f(x)$ is a valid pdf.

Solution

The constant k must satisfy two conditions:

$$\int_0^4 f(x)\,dx = 1$$

and

$$f(x) \ge 0; \quad 0 \le x \le 4$$

For Condition 1,

$$\int_0^4 \left(x^2 + kx + 3\right)dx = 1$$
$$= \frac{x^3}{3} + \frac{kx^2}{2} + 3x = 1$$
$$= \frac{64}{3} + 8k + 12 = 1$$

Solving for k yields,

$$k = -\frac{97}{24}$$

For Condition 2, determine $f(x)$ vs. x and ascertain whether $f(x)$ is positive over the range $0 \le x \le 4$. For $x = 3$

$$f(x) = 3^2 - \left(\frac{97}{24}\right)(3) + 3$$
$$= 9 - 12.2 + 3$$
$$= -0.3$$
$$= \text{negative value}$$

Since $f(x)$ is not positive for values of x between 0 and 4, the pdf provided by the engineer is not valid with the calculated value of k. A flunking grade may be in order for the intern.

APPLICATION 20.4

The auto-ignition temperature (AIT) is the temperature at which a substance will catch fire spontaneously in air, without a source of ignition. In the hydrocarbon series of chemicals, the AIT falls as the boiling point rises. The AIT of ethylene is about 480°C and that of heavy fuel oil about 250°C. Processing of these chemicals is often carried out at higher temperatures and any leaks that occur will ignite.

Some concern has arisen regarding the possibility of an explosion at a nanoparticulated plant where the nanoparticles are present in a hydrocarbon mixture. Preliminary studies indicate that the difference between the actual AIT and a calculated value in a particulate–vapor mixture is a random variable X (in °F) having the pdf

$$f(x) = 1.2e^{-1.2x}; \qquad x > 0$$
$$= 0; \qquad\qquad \text{elsewhere}$$

Calculate $P(2 < X < 6)$.

Solution

Applying the definition of cdf leads to

$$P(2 < X < 6) = \int_2^6 f(x)\,dx$$

Substituting and integrating yields

$$P(2 < X < 6) = \int_2^6 1.2e^{-1.2x}\,dx$$
$$= e^{-2.4} - e^{-7.2}$$
$$= 0.091 - 7.5 \times 10^{-4}$$
$$= 0.091 = 9.1\%$$

APPLICATION 20.5

An explosion reported at a nanochemical plant could have occurred as a result of three mutually exclusive causes involving nanoparticles: particle size distribution (PSD), particle concentration (PC), and particle shape (PS). It is estimated that such an explosion could occur with a probability of 0.75 as a result of PSD, 0.40 as a result of PC, and 0.30 as a result of PS. It is also estimated that the prior probabilities of the three possible causes are, respectively, 0.60, 0.30, and 0.10. What is the most likely cause of the explosion?

Solution

Represent the given event by B; that is, B is the explosion of the nanoparticles. Identify the antecedent events and represent them by A_1, A_2,..., A_n. Let A_1, A_2, A_3 denote, respectively, the events that PSD, PC, and PS are the causes of the explosion. The prior probabilities are as follows:

$$P(A_1) = 0.60$$
$$P(A_2) = 0.30$$
$$P(A_3) = 0.10$$

The conditional probabilities of the event B given each of the antecedent events as described in Equation 3.28 (in Chapter 3) are as follows:

$$P(B/A_1) = 0.75$$
$$P(B/A_2) = 0.40$$
$$P(B/A_3) = 0.30$$

Substitute these values in Equation 3.36 (in Chapter 3) to determine the revised (posterior) probabilities, that is,

$$P(A_1/B), P(A_2/B),...,P(A_n/B)$$

$$P(A_1/B) = \frac{(0.60)(0.75)}{(0.60)(0.75)+(0.30)(0.40)+(0.10)(0.30)}$$

$$= \frac{0.45}{0.45+0.12+0.03}$$

$$= 0.75$$

$$P(A_2/B) = 0.20$$
$$P(A_3/B) = 0.05$$

APPLICATION 20.6

If the probability of a valve employed at a nanoplant being defective is 15%, what is the probability of obtaining no defectives in a sample size of 2? Repeat the calculation for two defectives.

Solution

Employ the binomial distribution equation provided in Equation 5.2 (in Chapter 5) with $p=0.15$, $q=0.85$, and $n=2$.

$$P(X = x) = \frac{n!}{x!(2-x)!}(p)^x (q)^{2-x}$$

For $x=0$,

$$P(X = 0) = \frac{2!}{(0!)(2!)}(0.15)^0 (0.85)^2$$
$$= 0.7225 = 72.3\%$$

For exactly 2 defectives, $x=2$,

$$P(X = 2) = \frac{2!}{(2!)(0!)}(0.15)^2 (0.85)^0$$
$$= 0.0225 = 2.25\%$$

APPLICATION 20.7

The probability that each experiment in a nanoparticle laboratory will produce a "useable" chemical is 0.25. After 6 experiments, calculate the probability that a "useful" chemical will be realized

1. Exactly 2 times; and
2. More than 4 times

Solution

This involves application of the binomial distribution with $p=0.25$, $q=0.75$, and $n=6$. Apply the appropriate equation in Chapter 5.

1. For this case,

$$P(X = 2) = \frac{6!}{(2!)(4!)}(0.25)^2 (0.75)^4$$
$$= 0.297 = 30\%$$

2. For this case,

$$P(X > 4) = P(X = 5) + P(X = 6)$$

There are two terms to calculate,

$$P(X = 5) = \frac{6!}{(5!)(1!)}(0.25)^5(0.75)^1$$
$$= 0.0044$$
$$P(X = 6) = \frac{6!}{(6!)(0!)}(0.25)^6(0.75)^0$$
$$= 3.8 \times 10^{-6}$$

Thus,

$$P(X > 4) = 0.044 + 3.8 \times 10^{-6}$$
$$= 0.0044 = 0.44\%$$

As expected, the probability is low.

APPLICATION 20.8

Occupational mishaps are usually a function of three aspects of the work place: the human element, task variables (that is, the job itself), and the environmental element. The human element is briefly addressed as follows.

When evaluating human elements in relation to occupational mishaps, six factors generally must be taken into consideration.[1-4]

1. *Sex*: Within a given job setting, a worker's sex may increase the individual's propensity for accidents and injuries.
2. *Age*: Data seem to indicate that younger workers have a higher potential to become involved in accidents.
3. *Personality*: Personality factors such as tendency toward anger, discontent, excitability, and hostility, as well as a low order of adjustment and high impulsiveness can also contribute to maladjustments and accidents.
4. *Physical–physiological status*: A worker's physical and physiological capacity for work may have an impact on his or her accident and injury potential.
5. *Accident proneness*: The old notion that certain people are accident prone has been difficult to establish as fact.
6. *Education*: Education counts. A college education is preferred to a high school education.

Human error analysis is a procedure that evaluates the factors influencing the performance of operators, maintenance personnel, technicians, and so forth. Its purpose is to identify human errors and their accompanying effects. It has also been used to identify the cause of human errors. Kletz has an excellent treatment of this topic.[4]

A nanochemical company recently concluded that accident occurring at their plant could be attributed to any one of the earlier mentioned 6 human errors. If the probability associated with any accident is equality distributed between 6 causes, that is, $p = 1/6$, calculate the probability that for the past accidents that occurred, 2 were caused by (5) above, 2 were caused by (6) above, and each of the remaining 4 were caused by either (1), (2), (3), or (4).

Solution

Apply the multinomial equation.

$$P = P(1,1,1,1,2,2) = \frac{8!}{(1!)(1!)(1!)(1!)(2!)(2!)} \left(\frac{1}{6}\right)\left(\frac{1}{6}\right)\left(\frac{1}{6}\right)\left(\frac{1}{6}\right)\left(\frac{1}{6}\right)^2\left(\frac{1}{6}\right)^2$$

$$= 0.006 = 0.6\%$$

APPLICATION 20.9

The probability that an exposure to a nanocarcinogen at a site will be fatal is 0.80. Find the probability that 4 to 8 will die due to exposure for a group of 15 workers at the plant.

Solution

This involves the binomial distribution. See also Chapter 5 where it was noted that for this case,

$$P(4 \le X \le 8) = 1.0 - P(X \ge 9) - P(0 \le X \le 4)$$

One notes almost immediately that

$$P(0 \le X \le 4) \approx 0$$

Therefore, since P $(X \ge 9) = 0.982$,

$$P(4 \le X \le 8) = 1.0 - 0.982 - 0.0$$
$$= 0.018 = 1.8\%$$

APPLICATION 20.10

As described in Chapter 7, the hypergeometric distribution involves an extension of the binomial distribution. For this case, let a be the number of items in the category labeled "success." Then $N - a$ will be the number of items in the category labeled "failure." Let X denote the number of successes in a random sample of n items drawn without replacement from the set of N items. Then the random variable X has a hypergeometric distribution whose probability distributions function (pdf) is specified as follows (see also Chapter 7):

$$f(x) = \frac{\dfrac{a!}{x!(a-x)!} \dfrac{(N-a)!}{(n-x)!(N-a-n+x)!}}{\dfrac{N!}{n!(N-n)!}}; \quad x = 0,1,\ldots,\min(a,n) \quad (7.1)$$

The term $f(x)$ is the probability of x successes in a random sample of n items drawn without replacement from a set of N items, a of which are classified as successes and $N - a$ as failures. The term $\min(a, n)$ represents the smaller of the two numbers a and n, that is, $\min(a, n) = a$ if $a < n$ and $\min(a, n) = n$ if $n \le a$.

A sample of 5 nanochemicals is drawn at random *without replacement* from a lot of 1000, 5% of which are "contaminated." What is the probability that the sample contains exactly 2 contaminated chemicals?

Solution

The number of nanochemicals in the lot is 1000. Therefore, $N = 1000$. Once again associate "success" with drawing a contaminated nanochemical, and "failure" with drawing one that is not contaminated. Since 5 of the chemicals in the lot are contaminated, $a = 5$, and the problem states that a sample of 5 is drawn from the lot, that is, $n = 5$. Substitute the values of N, n, and a in the pdf of the hypergeometric distribution, that is,

$$f(x) = \frac{\dfrac{50!}{x!(50-x)!} \dfrac{950!}{(5-x)!(945+x)!}}{\dfrac{1000!}{(5!)(995)!}}; \quad x = 0,1,\ldots,5$$

One may now substitute the appropriate values of x to obtain the required probability.

$$P(\text{sample contains exactly 2 defectives}) = P(X = 2)$$

Therefore,

$$P(X = 2) = \frac{\dfrac{5!}{(2!)(3!)} \dfrac{95!}{(3)!(92!)}}{\dfrac{100!}{(5!)(95)!}}$$

$$= \frac{(10)(138,415)}{75,287,520}$$

$$= 0.0184 = 1.84\%$$

As expected, this is a very low probability.

APPLICATION 20.11

Ten percent of the chemicals produced in a certain nanomanufacturing process turn out to be unacceptable for further processing. Find the probability that in a sample of 10 chosen at random, exactly 2 will be discarded.

Solution

The probability of an unacceptable chemical is $p = 0.1$. Let X denote the number of unacceptable chemicals out of 10 selected. Assume that the Poisson distribution applies. For this case,

$$\lambda = np = (10)(0.1) = 1$$

Then, according to the Poisson distribution,

$$P(X = x) = \frac{\lambda^x e^{-\lambda}}{x!}$$

or

$$P(X = 2) = \frac{(1)^2 e^{-1}}{2!}$$
$$= 0.1839 = 18.4\%$$

APPLICATION 20.12

The probability that an individual will suffer a bad reaction from a nanochemical cream is 0.001 (fractional basis). Determine the probability that out of 2000 users

1. Exactly 3
2. More than 2

individuals will suffer a bad reaction.

Solution

Let X denote the number of individuals suffering a bad reaction. Since bad reactions are assumed to be rare events, one can assume that X is Poisson distributed, that is,

$$P(X = x) = \frac{\lambda^x e^{-\lambda}}{x!}$$

where

$$\lambda = np = (2000)(0.001) = 2$$

1. Here,

$$P(X = 3) = \frac{(2)^3 e^{-2}}{3!}$$

$$= 0.180 = 18.0\%$$

2. For this case,

$$P(X > 2) = 1 - \left[P(X = 0) + P(X = 1) + P(X = 2) \right]$$

$$= 1 - \left[\frac{(2)^0 e^{-2}}{0!} + \frac{(2)^1 e^{-2}}{1!} + \frac{(2)^2 e^{-2}}{2!} \right]$$

$$= 1 - 5e^{-2}$$

$$= 0.323 = 32.3\%$$

APPLICATION 20.13

The mean diameter of a sample of 200 nanoparticles produced by a machine is 0.502 μm and the standard deviation is 0.005 μm. The application of these particles is intended to allow a maximum tolerance in the diameter of 0.496–0.508 μm, otherwise the particles are considered unacceptable for use. Determine the percentage of unacceptable particles produced if the diameters are normally distributed.

Solution

Convert the data to the standard normal variable

$$Z_1 = (0.496 - 0.502)/0.005$$

$$= -1.2$$

$$Z_2 = (0.508 - 0.502)/0.005$$

$$= 1.2$$

The probabilities of acceptable particles, AP, is therefore

$$AP = (\text{area under normal curve between } Z_1 = -1.2 \text{ and } Z_2 = 1.2)$$

$$= (\text{twice the area between } Z = 0 \text{ and } Z = 1.2)$$

$$= 2(0.3849)$$

$$= 0.7698 = 77\%$$

Therefore, the percentage of unacceptable particles, UP, is

$$UP = 1.0 - 0.7698$$

$$= 0.2302 = 23\%$$

APPLICATION 20.14

The time to failure for a heat exchanger in a nanochemical facility is assumed to follow an exponential distribution with $\lambda = 0.15$ (per year). What is the probability of a failure within the first 2 years?

Solution

Refer to Chapter 11.

$$F(x) = \int_0^x \lambda e^{-\lambda x}\, dx = 1 - e^{-\lambda x}$$

For this case,

$$P(X \le 2) = \int_0^2 (0.15) e^{-(0.15)x}\, dx$$

$$= -\frac{0.15}{0.15} e^{-(0.15)x}\Big|_0^2$$

$$= e^{-0.3} + e^0$$

$$= 1 - e^{-0.3} = 1 - 0.741$$

$$= 0.259 = 25.9\%$$

Therefore, there is nearly a 26% probability that the exchanger will fail within the first 2 years.

APPLICATION 20.15

The following cumulative particle size data for a nanodischarge is provided in Table 20.1. Estimate the mass percent of particles[5–6]

1. Less than 1.0 μm, and
2. Greater than 10 μm

Solution

A plot of Table 20.1 (see Figure 20.1) on log-probability paper yields a straight line. Therefore, the distribution is log-normal.

1. To determine the percent of particles less than 1.0 μm, extrapolate the line to $d_p = 1.0$ μm and read the upper x axis. The answer is approximately 1%.
2. To determine the mass percent of particles greater than 10 μm (an old ambient air quality standard), read the x coordinate at 10 μm. The answer is approximately 50%.

TABLE 20.1
Particle Size Concentration Data

Particle Size Range (µm)	Concentration (µg/m³)	Weight in Size Range (%)	Cumulative % GTSS[a]
0–2	0.8	0.4	99.6
2–4	12.2	6.1	93.5
4–6	25.0	12.5	81.0
6–10	56	28	53.0
10–20	76	38	15.0
20–40	27	13.5	1.5
>40	3	1.5	—

[a] % GTSS represents the percent greater than stated size, where the stated size is the upper limit of the corresponding particle size range. Thus, 99.6% of the particles have a size equal to or greater than 2 µm.

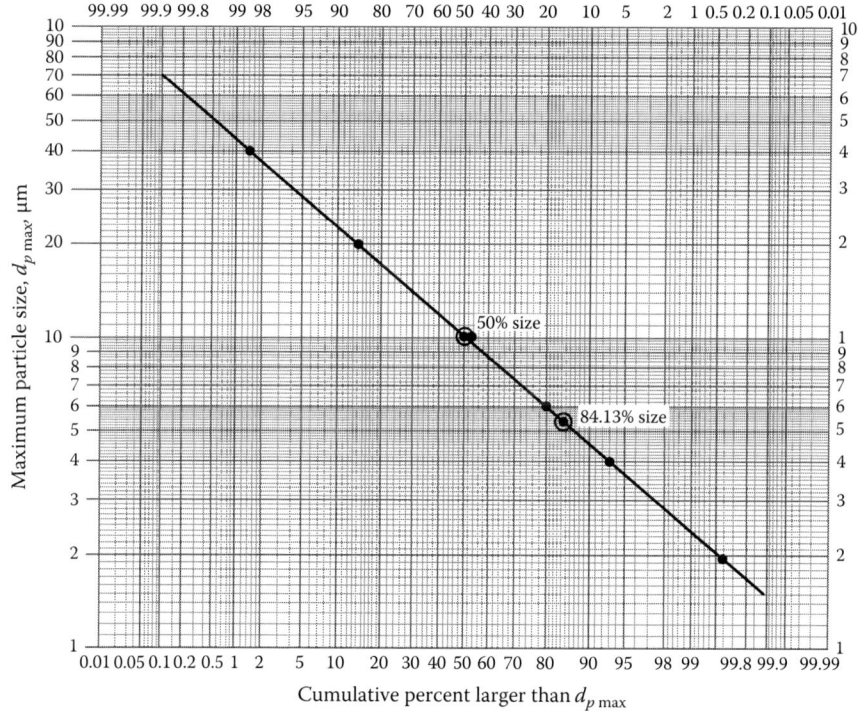

FIGURE 20.1 Log-probability distribution for data from Application 20.15.

APPLICATION 20.16

The probability that a pressure gauge in a Canadian nanofacility will not survive for more than 60 months is 0.95. How often should the gauge be replaced? Assume the time to failure is exponentially distributed and that the replacement time should be based on the gauge's expected life.

Solution

This requires the calculation of μ in the exponential model (see also Chapter 11) with units of (month)$^{-1}$. Once again,

$$F(t) = 1 - e^{-\lambda t}$$

Based on the information provided

$$P(T \le 60) = 0.95$$

or

$$0.95 = 1 - e^{-(\lambda)(1)}$$

Solving for λ gives

$$\lambda = -\frac{1}{60} \ln(1 - 0.95)$$
$$= 0.05$$

Because the expected time (or life), $E(T)$, is

$$E(T) = \frac{1}{0.05}$$
$$= 20 \text{ months}$$

The gauge should therefore be replaced in approximately 20 months.

APPLICATION 20.17

A particle size distribution (PSD) measurement device has a constant rate of failure of 0.005 per thousand hours. Find the probability that the device will operate for at least 25,000 h.

Solution

Substitute the failure rate

$$Z(t) = 0.005$$

into the describing equation in Chapter 12 which yields

$$f(t) = \exp\left[-\int_0^t 0.005 \, dt\right]$$
$$= 0.005e^{-0.005t}; \quad t > 0$$

as the pdf of T, the time to failure of the PSD device. Because t is measured in thousands of hours, the probability that the device will operate for at least 25,000 h is given by

$$P(T > 25) = \int_{25}^{\infty} -0.005e^{-0.005t} \, dt$$
$$= -e^{-\infty} + e^{-0.005(25)}$$
$$= 0 + e^{-0.125}$$
$$= 0.882 = 88.2\%$$

APPLICATION 20.18

A nanofacility employs a large number of thermometers of which 20% are of design I and 80% design II. Ten years later, there have been 59 thermometer failures of design I and 193 of design II. Comment on whether there is evidence of a difference between the two designs.

Solution

If there were no difference between the designs, the expected number of failures would be proportionally the same. For this application, first note that the total number of failures is $59 + 193 = 252$. Therefore,

$$\text{Failures, I} = (0.20)(252)$$
$$= 50.4$$
$$\text{Failures, B} = (0.80)(252)$$
$$= 201.6$$

Employ Equation 13.14 (in Chapter 13) to calculate χ^2:

$$\chi^2 = \frac{(59 - 50.4)^2}{50.4} + \frac{(193 - 201.8)^2}{201.8}$$
$$= 1.47 + 0.38$$
$$= 1.85$$

Because there are only two designs, there is only 1 degree of freedom. From Table 13.3 (in Chapter 13), one notes that with 1 degree of freedom, there is approximately only a 0.20 (20%) probability that χ^2 would be larger than 1.85. Therefore, the probability that there is no difference between the two thermometer designs is less than 20%.

APPLICATION 20.19

Theodore Associates has been requested to conduct a risk assessment at a chemical plant that is concerned with the consequences of two incidents that occur at approximately the same location in the plant and that are defined as follows[5]:

    I. An explosion resulting from the detonation of an unstable nanochemical.
    II. A continuous 240 g/s release of a resulting toxic chemical at an elevation of 125 m.

Two weather conditions are envisioned, namely, a northeast wind and a southwest wind with Stability Class B. Associated with these two wind directions are Events IIA and IIB, respectively, defined as follows:

    IIA—Toxic cloud to the southwest
    IIB—Toxic cloud to the northeast

Based on an extensive literature search, the probabilities and conditional probabilities of the occurrence of the defined events in any given year have been estimated by Dupont Consultants as follows:

$$P(\text{I}) = 10^{-6}$$
$$P(\text{II}) = 1/33,333$$
$$P(\text{IIA} \mid \text{II}) = 0.33$$
$$P(\text{IIB} \mid \text{II}) = 0.67$$

Note that $P(\text{IIA}|\text{II})$ represents the probability that Event IIA occurs *given* that Event II has occurred. The consequences of Events I, IIA, and IIB, in terms of number of people killed, are estimated as following:

    I—All persons within 200 m of the explosion center are killed; all persons beyond this distance are unaffected.
    IIA—All persons in a pie-shaped segment, 22.5° width (downwind of the source) are killed if the concentration of the toxic gas is above 0.33 µg/L; all persons outside this area are unaffected.
    IIB—Same as IIA.

Thirteen people are located within 200 miles of the explosion center but not in the pie-shaped segment described earlier. Eight people are located within the pie-shaped segment southwest of the discharge center; 5 are 350 miles downwind, three are 600 miles away at the plant fence (boundary). Another 6 people are located 500 miles away outside the pie-shaped segment but within the plant boundary. All individuals are at ground level.

Theodore Associates have been specifically requested to calculate the average annual individual risk (AAIR) based on the number of individuals potentially affected as well as the average risk based on all other individuals within the plant boundary. *Hint*: Perform atmospheric dispersion calculations at various distances from the emission source and combine these predicted concentrations with consequence information from the problem statement.

Solution

Draw a line diagram of the plant layout and insert all pertinent data and information (see Figure 20.2). An event tree[1–3] for the process is presented in Figure 20.3.

First, calculate the probability of Event IIA occurring; also calculate the probability of Event IIB occurring as follows:

$$P(\text{IIA}) = P(\text{II})P(\text{IIA}|\text{II}) = \left(\frac{1}{33,333}\right)(0.33) = \frac{1}{100,000} = 10^{-5}$$

$$P(\text{IIB}) = P(\text{II})P(\text{IIB}|\text{II}) = \left(\frac{2}{33,333}\right)(0.67) = \frac{2}{100,000} = 2\times10^{-5}$$

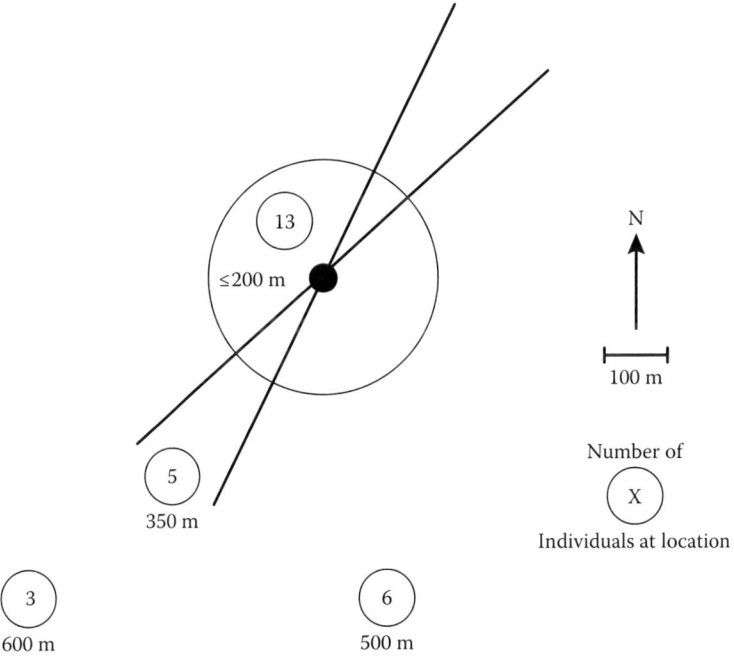

FIGURE 20.2   Plant layout and pertinent data for Illustrative Example 20.19.

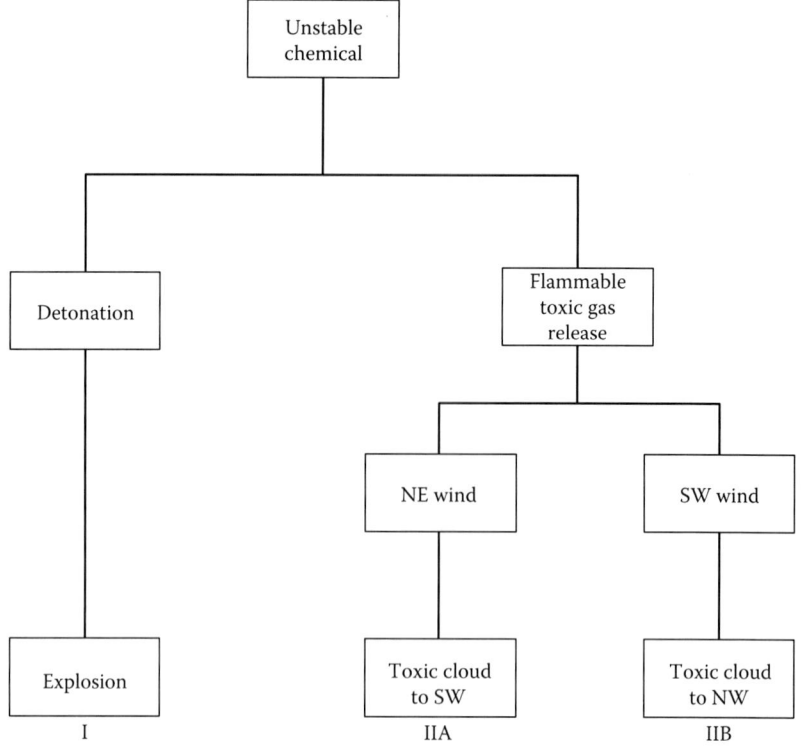

FIGURE 20.3 Event tree for Application 20.19.

Perform a dispersion calculation to determine the zones where the concentration of the chemical exceeds 0.33 µg/L. Assume a continuous emission for a point source.[6] To maintain consistent units, convert wind speed mph to m/s and concentration from µg/L to g/m³ as follows:

$$u = (6.0 \text{ miles/h})(5280 \text{ ft/mile})(1 \text{ h/3600 s})(0.3048 \text{ m/ft}) = 2.68 \text{ m/s}$$
$$c = (0.33 \text{ µg/L})(1 \text{ g/10}^6 \text{ µg})(10^3 \text{ L/m}^3) = 3.3 \times 10^{-4} \text{ g/m}^3$$

These values are then used as input to the Pasquill–Gifford model[6] for centerline, ground-level concentrations of a continuous source of pollutant at an elevated emission height of H^* as shown in the following:

$$c(x,0,0;125) = \frac{q}{\pi \sigma_y \sigma_z u} \left\{ \exp\left[-\frac{1}{2}\left(\frac{H^*}{\sigma_z}\right)^2 \right] \right\}$$

$$= \left(\frac{240 \text{ g/s}}{\pi (\sigma_y)(\sigma_z) 2.68 \text{ m/s}} \right) \left\{ \exp\left[-\frac{1}{2}\left(\frac{125 \text{ m}}{\sigma_z}\right)^2 \right] \right\}$$

TABLE 20.2

Downwind Concentration Profile

x (m)	σ_y(m)	σ_z(m)	c (g/m³)
300	47	30	3.43×10^{-6}
400	60	41	1.11×10^{-4}
500	75	52	4.07×10^{-4}
550	80	60	6.78×10^{-4}
600	90	65	7.67×10^{-4}
700	105	77	9.44×10^{-4}
800	120	90	1.01×10^{-4}
900	150	110	9.06×10^{-4}
1000	170	140	8.04×10^{-4}
1500	250	240	4.15×10^{-4}
1700	275	275	3.40×10^{-4}
2000	300	380	2.37×10^{-4}

The downwind concentrations can be calculated based on the above equation. A linear interpolation indicates that the maximum ground level concentration (GLC) is approximately 1.01×10^{-3} g/m³ and is located at a downwind distance of about 800 m. In addition, the "critical" zone, where the concentration is above 3.3×10^{-4} g/m³, is located between 475 and 1800 m. The concentration results for select downwind distances are provided in Table 20.2.

Now determine which individuals within the pie-shaped segment downwind from the source will be killed if either accident (I or II) occurs. Referring to Figure 20.2, 13 individuals within the 200 m radius will die from Accident I. Three individuals located in the pie-shaped segment and 600 m southwest of the emission source will die from Accident II. The five individuals located 350 m southwest of the emission source are in the path of the dispersing plume but are all outside the critical zone. The six individuals located outside of the pie-shaped impact segment are within the plant boundary but are not potentially affect by either the explosion or the dispersing plume.

The total amount of deaths (TAD) for the process if the accident occurs are therefore

$$TAD = 13 + 3 = 16 \text{ deaths/year}$$

The total annual risk (TAR) is obtained by multiplying the number of people in each impact zone by the probability of the event affecting that zone and summing the results. Thus,

$$TAR = (13)P(I) + (3)P(IIA) = (13)(10^{-6}) + (3)(10^{-5}) = 4.3 \times 10^{-5}$$

The AAIR is obtained by dividing this result by number of people in the impact zone. The average annual risk (AAR) is calculated based only on the "potentially affected" people. Since 21 people are "potentially affected," that is, are within the impact area of the explosion or are in the path of the dispersing plume, the AAR is determined to be as follows:

$$AAR = \frac{4.3 \times 10^{-5}}{21} = 2.05 \times 10^{-6}$$

The AAIR is based on all the individuals within the plant boundary. For this case study, this AAIR is now based on 27 rather than 21 individuals. Thus,

$$AAIR = \frac{4.3 \times 10^{-5}}{27} = 1.6 \times 10^{-6}$$

It should be noted that only one "average" weather condition was considered in the example. However, one often selects the worst-case weather condition that corresponds with a reasonable probability of occurrence in the location of the site being evaluated. Employing this worst-case condition produces risk results on the conservative side. An analysis that includes a full spectrum of wind speeds, directions, and stability classes would obviously provide a more complete set of risk assessment calculations than is provided here.

REFERENCES

1. L. Theodore and F. Taylor, *Probability and Statistics*, Theodore Tutorials (originally published by USEPA, RTP, NC), East Williston, NY, 1996.
2. S. Shaefer and L. Theodore, *Probability and Statistics Applications in Environmental Science*, CRC Press/Taylor & Francis Group, Boca Raton, FL, 2007.
3. L. Theodore and R. Dupont, *Environmental Health and Hazard Risk Assessment: Principles and Calculations*, CRC Press/Taylor & Francis Group, Boca Raton, FL, 2013.
4. T. Kletz, *What Went Wrong? Case Histories of Process Plant Disasters*, Gulf Publishing, Houston, TX, 1985.
5. D. Hendershot, A simple example problem illustrating the methodology of chemical process quantitative risk assessment, paper presented at *AIChE Mid-Atlantic Region Day in Industry for Chemical Engineering Faculty*, Bristol, PA, April 15, 1988.
6. L. Theodore, *Air Pollution Control Equipment Calculations*, John Wiley & Sons, Hoboken, NJ, 2008.

Index